EAI/Springer Innovations in Communication and Computing

Series editor
Imrich Chlamtac, European Alliance for Innovation, Gent, Belgium

Editor's Note

The impact of information technologies is creating a new world yet not fully understood. The extent and speed of economic, life style and social changes already perceived in everyday life is hard to estimate without understanding the technological driving forces behind it. This series presents contributed volumes featuring the latest research and development in the various information engineering technologies that play a key role in this process.

The range of topics, focusing primarily on communications and computing engineering include, but are not limited to, wireless networks; mobile communication; design and learning; gaming; interaction; e-health and pervasive healthcare; energy management; smart grids; internet of things; cognitive radio networks; computation; cloud computing; ubiquitous connectivity, and in mode general smart living, smart cities, Internet of Things and more. The series publishes a combination of expanded papers selected from hosted and sponsored European Alliance for Innovation (EAI) conferences that present cutting edge, global research as well as provide new perspectives on traditional related engineering fields. This content, complemented with open calls for contribution of book titles and individual chapters, together maintain Springer's and EAI's high standards of academic excellence. The audience for the books consists of researchers, industry professionals, advanced level students as well as practitioners in related fields of activity include information and communication specialists, security experts, economists, urban planners, doctors, and in general representatives in all those walks of life affected ad contributing to the information revolution.

About EAI

EAI is a grassroots member organization initiated through cooperation between businesses, public, private and government organizations to address the global challenges of Europe's future competitiveness and link the European Research community with its counterparts around the globe. EAI reaches out to hundreds of thousands of individual subscribers on all continents and collaborates with an institutional member base including Fortune 500 companies, government organizations, and educational institutions, provide a free research and innovation platform.

Through its open free membership model EAI promotes a new research and innovation culture based on collaboration, connectivity and recognition of excellence by community.

More information about this series at http://www.springer.com/series/15427

Baoliu Ye • Weihua Zhuang • Song Guo
Editors

2nd International Conference on 5G for Ubiquitous Connectivity

5GU 2018

Springer

EAI
RESEARCH MEETS INNOVATION

Editors
Baoliu Ye
National Key Laboratory for Novel
Software Technology
Nanjing University
Nanjing, China

Weihua Zhuang
Department of Electrical and Computer
Engineering
University of Waterloo
Waterloo, ON, Canada

Song Guo
Department of Computing
The University of Polytechnic University
Kowloon
Hong Kong, Kowloon, Hong Kong

ISSN 2522-8595 ISSN 2522-8609 (electronic)
EAI/Springer Innovations in Communication and Computing
ISBN 978-3-030-22315-1 ISBN 978-3-030-22316-8 (eBook)
https://doi.org/10.1007/978-3-030-22316-8

This Springer imprint is published by the registered company Springer Nature Switzerland AG.
The registered company address is: Gewerbestrasse 11, 6330 Cham, Switzerland

Preface

We are delighted to introduce the proceedings of the 2nd International Conference on 5G for Ubiquitous Connectivity (5GU 2018). The aim of this conference is to bring together researchers and developers as well as regulators and policy makers to present their latest views on 5G: New networking, new wireless communications, resource control and management, future access techniques, new emerging applications, and of course, latest findings in key research activities on 5G.

The technical program of 5GU 2018 consisted of 15 full papers at the main conference tracks. The conference tracks were Track 1—New networking for 5G and beyond; Track 2—New wireless communications for 5G; Track 3—Resource control and management for 5G; Track 4—Future access techniques, and Track 5—New emerging applications. Aside from the high-quality technical paper presentations, the technical program also featured two keynote speeches. The two keynote speeches were by Dr. Ing. Thorsten Herfet from Saarland Informatics Campus, Germany, and Dr. Shi Jin from Southeast University, China.

Coordination with the steering chair, Prof. Imrich Chlamtac, was essential for the success of the conference. We sincerely appreciate the contribution of two general chairs, Prof. Baoliu Ye and Prof. Weihua Zhuang. It was also a great pleasure to work with such an excellent organizing committee team for their hard work in organizing and supporting the conference. In particular, the Technical Program Committee, led by our TPC Chair, Prof. Song Guo, who have completed the peer review process of technical papers and made a high-quality technical program. We are also grateful to all the authors who submitted their papers to the 5GU conference.

We strongly believe that 5GU 2018 conference provides a good forum for all researcher, developers, and practitioners to discuss all science and technology aspects that are relevant to 5G. We also expect that the future 5GU conference will be as successful and stimulating as indicated by the contributions presented in this volume.

Nanjing, China — Baoliu Ye
Waterloo, ON, Canada — Weihua Zhuang
Hong Kong, Kowloon, Hong Kong — Song Guo
Aizuwakamatsu, Japan — Peng Li

Conference Organization

Steering Committee	
Imrich Chlamtac	University of Trento, Italy
Organizing Committee	
General Chairs	
Baoliu Ye	Nanjing University, China
Weihua Zhuang	University of Waterloo, Canada
TPC Chair	
Song Guo	Hong Kong Polytechnic University
Local Chair	
Xin Wang	Hohai University, China
Workshops Chair	
Hongzi Zhu	Shanghai Jiaotong University, China
Publicity & Social Media Chair	
Guoping Tan	Hohai University, China
Publications Chair	
Peng Li	The University of Aizu, Japan
Web Chair	
Xujie Li	Hohai University, China
Conference Manager	
Kristina Lappyova	EAI
Technical Program Committee	
Shravan Garlapati	Virginia Tech University
Xiaojun Hei	Huazhong University of Science and Technology, China
Peng Liu	Hangzhou Dianzi University, China
Shengli Pan	China University of Geosciences (Wuhan), China
Tian Wang	Huaqiao University, China
Xiaoyan Wang	Ibaraki University, Japan
Xiaobo Zhou	Tianjin University, China
Shigeng Zhang	Central South University, China

Contents

Collaborative Inference for Mobile Deep Learning Applications

Qinglin Yang, Xiaofei Luo, Peng Li, and Toshiaki Miyazaki

1 Introduction

Algorithmic breakthroughs of deep learning in the past decades has attracted wide interest of developing artificial intelligence (AI) empowered mobile applications, such as Tencent QQ, Google Map, Apple Health, and Avast Mobile Security, etc., to conduct language translation, object recognition, health monitoring, and malware detection. The intelligent services provided by these mobile applications generally enable people to enjoy a more convenient as well as smarter mobile life. Although today's mobile devices become much more powerful than ever with greater computing capability and longer battery life, it might notice that not every people is able to be equipped with the newest and most powerful mobile devices. This indicates that significant heterogeneity (of available storage, CPUs, and batteries) exist between peoples' mobile devices. Furthermore, such heterogeneity will also emerge due to the different preferences of how people to use mobile devices, and sometimes leads related services to interrupt. It is an interesting yet much challenging topic to keep the accessibility of mobile services.

A nature way to tackle this challenge is to employ cloud computing by offloading the computation tasks to remote servers (aka on the cloud). For example, when the local mobile device needs to recognize the man in a picture, it only needs to upload this picture to the cloud and waits for a remote response of the final recognition result. However, there are two major concerns about this kind of could-computing based method: The first is the data transmission will consume a great amount of bandwidth for the cloud side. The traffic loads will get heavier as the users

Q. Yang · X. Luo · P. Li (✉) · T. Miyazaki
School of Computer Science and Engineering, University of Aizu, Aizuwakamatsu, Japan
e-mail: d8192105@u-aizu.ac.jp; d8202105@u-aizu.ac.jp; pengli@u-aizu.ac.jp; miyazaki@u-aizu.ac.jp

B. Ye et al. (eds.), *2nd International Conference on 5G for Ubiquitous Connectivity*, EAI/Springer Innovations in Communication and Computing,
https://doi.org/10.1007/978-3-030-22316-8_1

accumulate, and eventually make adverse impact on the cloud's QoS; The second concern is the transmission latency between the local mobile device and the remote cloud. In some emergency scenarios, users might expect near real-time response from the remote, while the transmission latency will be a big problem. To address these challenges, local offloading like fog computing and mobile edge computing is developed. So that many mobile devices now are able to contribute great GPU or even NPU computation capability. These mobile devices can partially serve the role of the cloud computation, and is fast to connect through Wi-Fi, Bluetooth, or near field communication(NFC).

In this paper, we propose employ local offloading to enable collaborative inference among local mobile devices. We first use a random structure to model the connections among mobile devices regarding their mobility. After the local link connections between mobile devices are established, the transmission latency on each link is assumed to remain constant yet various from each other. In what we show later the practical inference procedure is near real-time, mobile devices therefore are reasonably regarded as staying static until they receive the computation results. Then to accurately select the best local mobile device as the computation node illustrated in Fig. 1, our main concern naturally focuses on minimizing the whole time costs, which are induced by the data/result transmission between the computation nodes and the user nodes (that offload computation tasks), and task computation in the computation nodes. Unfortunately, the local connections will be updated from time to time due to the mobility of local mobile devices, making the optimal selection of computation nodes at one time not always suitable to the next time. This requires our collaborative inference scheme to not only find the optimal set of computation nodes in a short time, but also to be able to track the optimal

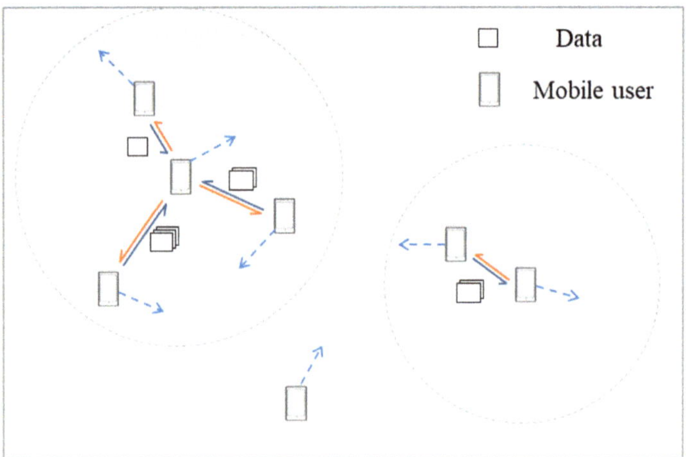

Fig. 1 System overview

selections in a dynamic environment. Obviously, optimization methods that are capable of constantly adapting the solution to a changing environment are expected. To meet this demand, we propose to employ partial swarm optimization (PSO) that is a versatile population-based stochastic optimization technique, to help design our collaborative inference scheme. The contributions of our work are summarized in the followings:

- We propose a system model with random structures to describe the (locally) collaborative inference among mobile services;
- We design an algorithm based on PSO to efficiently minimize the total time costs for collaborative inference, with a dynamic procedure of selecting the optimal computing nodes from local mobile devices;
- We conduct extensive simulations to evaluate the performances of our proposed algorithm, and demonstrate its comprehensive advantages to the optimal results obtained from Gurobi.

The rest of this paper are organized as below: in Sect. 2, we introduce the motivation of our work; in Sect. 3, we firstly formulate our system model and then we detail the PSO algorithm which is used to solve our problem; and the corresponding evaluation results are presented in Sect. 4. We demonstrate the related works in Sect. 5, and finally conclude this paper and in Sect. 6.

2 Motivation

We conduct experiments using three typical CNN models (AlexNet [10], Goog LeNet [13], and Lenet [11]) using Nvidia GTX1080. We collect GPU runtime information of the inference time by using Linux shell command. The CNN models are trained by Caffe [8], a popular open-source conventional neural networks framework which is widely used in both academia and industry. The training process and inference data of Lenet come from caffe models. As for GoogLeNet and AlexNet, we construct two figure recognition models with 209 classes in order to keep same model size with Lenet. The three models have the same amount of inference data(10000) downloaded from Imagenet [4]. And then, inference time under different batch sizes as shown in Fig. 2. The lines in Fig. 2 represents inference time per image on each architecture, with a function of image batch size(from 1 to 512). We notice that inference time across different batch sizes with a logarithmic ordinate. Missing data points are due to lack of enough graphics memory required to process larger batches. The inference time costing gradually decrease as the increase of batch size. Motivated by the trend, We think that to collect multi-users' data to handling by paralleling will be better than individually, though the transmission delay should be considered. We also find the view is novel and meaningful to research. We will introduce the model about the view detailed in next section.

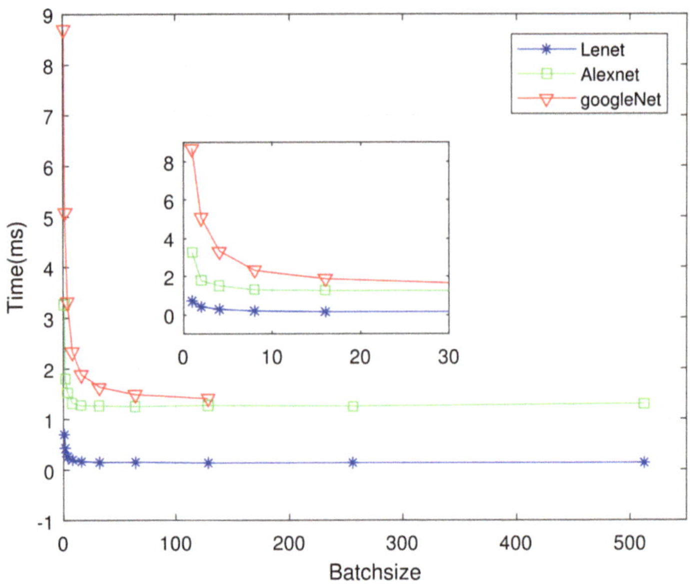

Fig. 2 Inference time *vs.* Batch size. This chart shows inference time across different batch size. Missing data points are due to lack of enough graphic memory required to process larger batches

3 System Model

We consider a set N of mobile devices running an application powered by deep neural networks (DNNs). The DNN model has been well trained and installed on mobile devices. Each device $i \in N$ has an amount of n_j data to process using equipped mobile GPU. These devices can connect with each other using direct links, e.g., Bluetooth and WiFi Direct, or cellular links. The neighbors of node i is included in set N_i. The communication delay between two devices i and j is denoted by d_{ij}. As we have shown that ML workload batching on GPU can effectively reduce the processing time, multiple mobile devices can aggregate their workloads on a single one.

We define a binary variable x_i to indicates whether node $i \in N$ is an aggregation node.

$$x_i = \begin{cases} 1, & \text{aggregation node} \\ 0, & \text{otherwise.} \end{cases}$$

Note that some nodes process only their own data, without receiving workloads from other nodes. We also treat them as aggregation nodes with $x_i = 1$. Each non-aggregation node may connect to multiple aggregation nodes. We define a variable y_{ij} to indicate the portion of workloads offloaded from node i to j. Since aggregation nodes do not offload their workload to others, we have $\sum_{j \in N_i} y_{ij} = 0$. For each non-aggregation node, we have $\sum_{j \in N_i} y_{ij} = 1$. In summary,

$$y_{ij} \leq x_i, \forall i \in N, j \in N; \tag{1}$$

$$x_i + \sum_{j \in N} y_{ij} = 1, \forall i \in N. \tag{2}$$

We define a total cost T_i of node i as the sum of its computation and communication cost, i.e.,

$$T_i = 2 * \sum_j d_{ij} * y_{ij} + f\Big(\sum_j y_{ji} * n_j\Big), \tag{3}$$

where $f(\cdot)$ is a non-decreasing function that describes the relationship between GPU processing time and workloads. With an objective of minimizing the total of cost among all mobile devices, our studied problem can be formulated as:

$$\min \sum_{i \in N} T_i$$

subject to: (1), (2) and (3).

In this section, we propose a heuristic algorithm based on particle swarm optimization (PSO), which is motivated by the phenomenon of bird predating. The key of PSO is to iterative improve a candidate solution with regard to a given measure of quality [12]. The solution obtained by the PSO may not be a theoretically optimal, but it can quickly generate a solution with satisfied performance in practice.

3.1 Procedure for PSO

The Particle Swarm Optimization procedure consists of $\boldsymbol{v_i}$ and $\boldsymbol{X_i}$ two parts' update, as shown in line 6 of algorithm and line 7 of algorithm. The $\boldsymbol{v_i}$ is a group of randomly generated feasible break-reconnect information which consists of n_i, m_i and r_i, represented by (4).

$$[n_i, m_i, r_i] \in \boldsymbol{v_i} \tag{4}$$

The n_i denotes the non-aggregation node which should removed from m_i. The m_i denotes the aggregation node which connected to n_i. The r_i denotes the other aggregation node. A simple instance shown in Fig. 3 is used to explain the update for $\boldsymbol{X}$ in line 6 of algorithm, in which the structure of $\boldsymbol{v_i}$ and $\boldsymbol{X_i}$ are provided. In

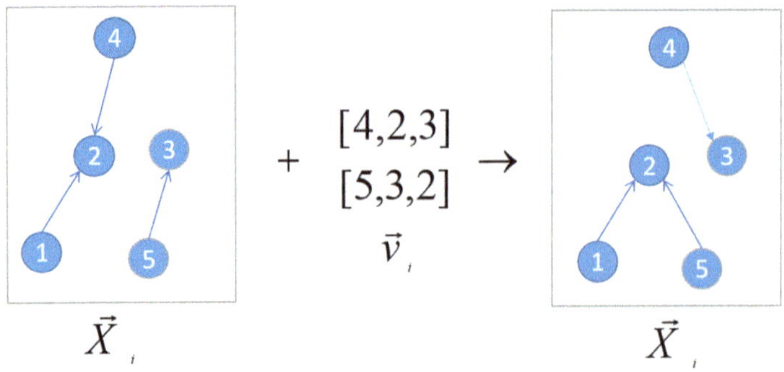

Fig. 3 Simple instance for line 7 of algorithm

Table 1 Variables and symbols

Notations	Description
$\boldsymbol{X_i}$	The present solution
$\boldsymbol{v_i}$	A group of randomly generated feasible break-reconnect information
$\boldsymbol{p_d}$	The local optimum in dth loop
p_g	The global optimum in all previous $\boldsymbol{p_d}$
w	The inertia weight
c_1, c_2	Acceleration constants, which are used to adjust step
h_1, h_2	Two random functions, whose field is [0,1], which are used to increase search randomness
$\boldsymbol{V}$	The set of $\boldsymbol{v_i}$
n_i	The non-aggregation node which should removed from m_i
m_i	Denotes the aggregation node which connected to n_i
r_i	Denotes the other aggregation node
F	The fitness for swarm
f	Describes the relationship between GPU processing time and workloads
P	Particle swarm, it is equal to the dimensions of $\boldsymbol{V}$

addition, R in line 6 of algorithm represents a set operations generate randomly in each iteration. The other major notations used in this algorithm are summarized in Table 1.

For instance, We assume there are five mobile nodes in the model, which may construct many different typologies as solutions for information propagation. We use $\boldsymbol{X_i}$ to represent the ith solution, which shown in the left of Fig. 3. The ith solution has two aggregation nodes whose number is 2 and 3 with other non-aggregation nodes connected. In order to get the optimal solution, the algorithm gives a operation set to make the ith solution converge to optimal solution, that is giving it a 'velocity' v_i towards the new solution.

The diagram shows the instance v_i as [4, 2, 3] and [5, 3, 2], in which the first set means the operation for non-aggregation node 4 would break the connectivity with aggregation node 2, and create a new connectivity to aggregation node 3; the similar operation towards the latter set. We named the aforementioned operation as *break-reconnect*. After the operation, the ith solution gets update, as shown in the right of Fig. 3.

As for a large model, there are much more complicated conditions that includes isolated aggregation node without nodes connected. There is also the condition for solution that makes aggregation node convert into non-aggregation,and then reconnect to other aggregation node. Hence, the previous non-aggregation nodes need to reallocate.

Line 6 of algorithm denotes how to get new break-reconnect information, and line 7 of algorithm represents how to get new solution. The right side in line 6 of algorithm is comprised of break-reconnect information, local optimum and global optimum. Then, according to update $\boldsymbol{v_i}$, the $\boldsymbol{X_i}$ gets its update. Parameters c_1 and c_2 are used to adjust the maximum step of iteration. In addition, h_1 and h_2 are two random numbers which contributed to search randomness.

3.2 Description of the Algorithm

First, in the initialization step, input $V = \{\boldsymbol{v_1}, \boldsymbol{v_2}, \ldots, \boldsymbol{v_p}\}$ is a set of $\boldsymbol{v_i}$. The operation for particle made by the break-reconnect information $\boldsymbol{v_i}$ is to remove the connectivity between the n_i and m_i, and then create a new connectivity for n_i with the aggregation node r_i. P is particle swarm, which is equal to the dimensions of V. After input V and P, we initialize $\boldsymbol{p_d}$ as p_0 which denotes the local optimum related to best solution in P before the loop start.

Then, we should consider all kinds of structures of $\boldsymbol{v_i}$. Firstly, we should consider when the n_i transformed from general node to a aggregation node, the n_i need to disconnect the connectivity without establishing any connection with others aggregation nodes at the time. Under the situation, we set $c_i = 0$. Similarly, there is also a case where r_i in $\boldsymbol{v_i}$, which means that the n_i who is the aggregation node in the old particle should be converted into the non-aggregation node in new particle and establish connectivity with aggregation node m_i. At this time, if the n_i acted as aggregation node in old particle has no connectivity with other non-aggregation nodes, then we just create a connection to aggregation node m_i in our algorithm.

When go in the loop, m_i, r_i, $\boldsymbol{v_i}$ and $\boldsymbol{X_i}$ are updated in each iteration. In addition, the fitness(k) can be calculated according to (3) and $f(\cdot)$. Once get the fitness value F, the local optimum and global optimum can be determined by line 1 of algorithm and line 15 of algorithm.

If the n_i is selected as the aggregation node by other general nodes in the old particle, we need to look for the computation which meet the condition that can be connected with other nodes in new particles. In addition, through analysis, it can be found that m_i and r_i are not equal to 0 at the same time. Finally, the output p_g is the global optimum we seek to.

Algorithm 1: Implementation of PSO algorithm

Input: V, P;
$d = 1$;
while $d \leq Loop$ **do**
  for k in P **do**
    for i in N **do**
      $\boldsymbol{v_i} \leftarrow w * \boldsymbol{R} + c_1 h_1 \otimes (\boldsymbol{p_d} - \boldsymbol{X_i}) + c_2 h_2 \otimes (p_g - \boldsymbol{X_i})$;
      $\boldsymbol{X_i} \leftarrow \boldsymbol{X_i} + \boldsymbol{v_i}$;
    end for
    $F \leftarrow Fitness(k)$;
    if $\boldsymbol{p_d} > F$ **then**
      $\boldsymbol{p_d} \leftarrow F$;
    end if
  end for
  if $p_g > \boldsymbol{p_d}$ **then**
    $p_g \leftarrow \boldsymbol{p_d}$;
  end if
  $d = d + 1$;
end while
Output: p_g;

4 Performance Evaluation

4.1 Settings

In this subsection, we implement an extensive simulation to evaluate the performance of PSO algorithm. We compare our approach with the optimal solution implemented by Gurobi that is state-of-the-art (http://www.gurobi.com/products/features-benefits). In the evaluation, assuming there are no more than 30 users in our experiment environment, because of limitation of the license of Gurobi. The algorithms write in Python, and the program runs on DELL, whose CPU Core is i5@2.30GHZ and memory 16GB. At the beginning, we set the popsize as 30 for our PSO. In other words, there are 30 groups random solution initially. The weight w and $Loop$ are normally recommended as 0.5 and 50.

4.2 Results Prediction

In this subsection, we show the experimental results for overall prediction performance of the proposed model. In Fig. 4, we mainly compare three conditions Random, PSO, and Optimal. Considering the Random, whose fitness calculated with

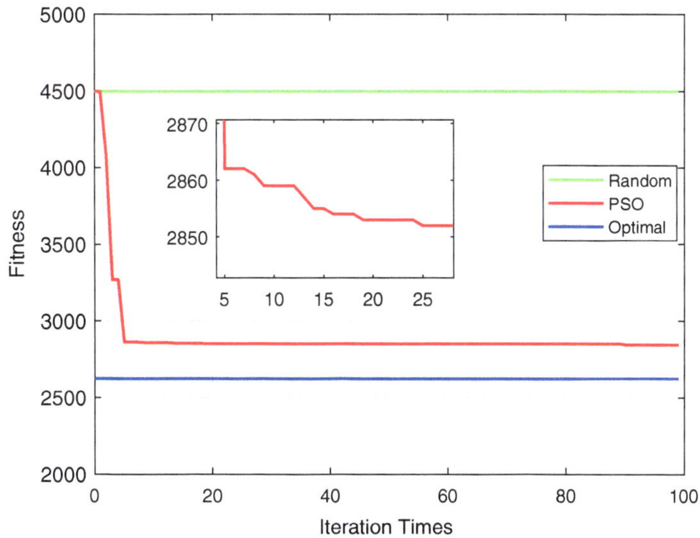

Fig. 4 Performance analysis comparison of the proposed algorithm with Gurobi and all users act as aggregation node

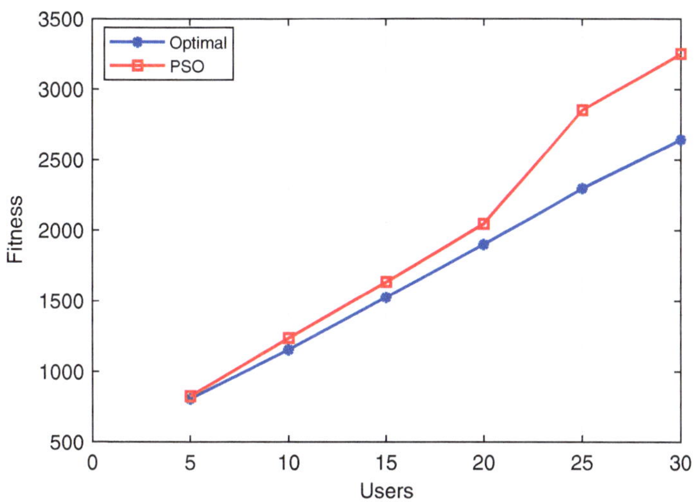

Fig. 5 With the increment of the magnitude of users N, which ranging from 1 to 30, the fitness of Optimal and PSO maintain the same growth trend

the condition that each user is treated as aggregation node. The red line denotes the optimal fitness generated by Gurobi, with 30 users in the environment. The green curve represents the process of iteration of fitness for our algorithm. As shown, when the iteration times exceeds 30, the green line tends to converge, whose value is far smaller than Random and close to the optimal. In Fig. 5, with the increase of

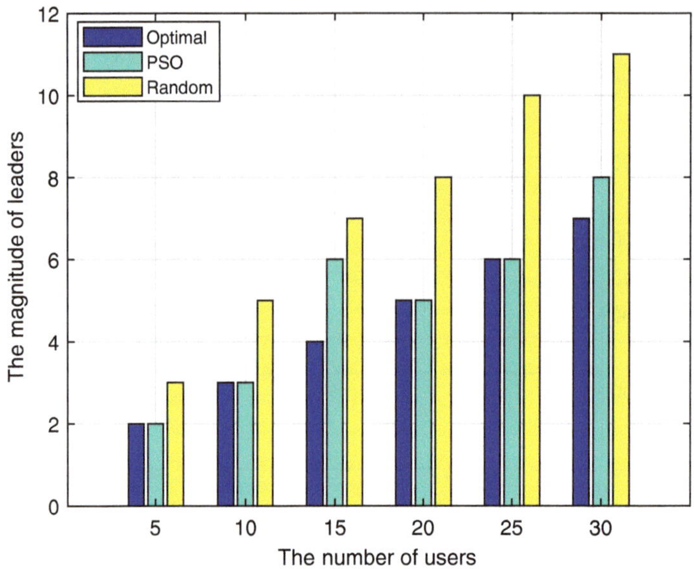

Fig. 6 The magnitude of aggregation nodes selected by different methods

the magnitude of users *N*, which range from 1 to 30, the fitness of Optimal and PSO maintain the same growth trend. And the value of fitness gained by PSO is very close to Optimal. In Fig. 6, the number of aggregation nodes of Random generated randomly at the beginning of initialization. Compared with Random, the number of aggregation nodes of Optimal and PSO is always smaller.

5 Related Work

5.1 Inference Process in Machine Learning

Much research work about efficiency in machine learning and cooperation in the mobile cloud has been done. In [3], Sharan Chetlur et al. improve the performance by 36% for convolution neural networks on caffe framework and reduce the memory consumption. In [1], Alfredo et al. do any analysis on accuracy, power consumption, inference time, memory footprint by experiments on framework named caffe. In [6], Han et al. benchmark the layer-wise speed up on CPU, GPU, and mobile GPU by deep compression for networks. In [14], Tang et al. propose a client-architecture where training process is implemented in a server, and then mobile device download the trained predictor from the server to make transmission decisions.

5.2 *Job Scheduling and Cooperation*

In [2], the lab established a SmartLab with 40 Android devices which cloud provide an open testbed to facilitate research and smart phone applications can be deployed massively. In [7], Heyi et al. propose a back-end general architecture which is able to require crowd-sourcing for mobile applications. And they adopt Microsoft Azure cloud computing platform to deploy their back-end. In [15], Yao et al. introduce a mobile cloud service framework based on crowdsourcing which meets mobile users requirement by sensing their context information and provide corresponding services to each of the users. In [9], Considering a dynamic network in which mobile devices may join and leave the network at any time, Ke et al. merge crowdsourcing into existing mobile cloud framework where data acquisition and processing can be conducted. In [5], Fan et al. propose a novel privacy-aware and trustworthy data aggregation protocol based on the malicious behavior like submitting data to damage the fog system for mobile sensing.

6 Conclusions

In this paper, we construct a model for transmission to implement local cooperation among mobile devices motivated by the inference process of machine learning, which can be used to guide the mobile users to choose which approach to handle their data for the goal of improving efficiency. By performance evaluation, we find that the collaborative inference scheme can reduce global dealing time in given field compared with handling the data which is affected by the high transmission latency between mobile device and cloud. As a global optimization random search algorithm, the particle swarm optimization algorithm has the characteristics of fast convergence and high precision.

References

1. Canziani, A., Paszke, A., Culurciello, E.: An analysis of deep neural network models for practical applications (2016). Preprint. arXiv:1605.07678
2. Chatzimilioudis, G., et al.: Crowdsourcing with smartphones. IEEE Internet Comput. **16**(5), 36–44 (2012)
3. Chetlur, S., et al.: cuDNN: efficient primitives for deep learning (2014). Preprint. arXiv:1410.0759
4. Deng, J., Dong, W., Socher, R., Li, L.-J., Li, K., Fei-Fei, L.: ImageNet: a large-scale hierarchical image database.In: IEEE Conference on Computer Vision and Pattern Recognition (2009)
5. Fan, J., Li, Q., Cao, G.: Privacy-aware and trustworthy data aggregation in mobile sensing. 2015 IEEE Conference on Communications and Network Security (CNS). IEEE, Piscataway (2015)

6. Han, S., Mao, H., Dally, W.J.: Deep compression: compressing deep neural networks with pruning, trained quantization and huffman coding (2015). Preprint. arXiv:1510.00149
7. Heyi, M.H., Rossi, C.: On the evaluation of cloud web services for crowdsourcing mobile applications. In: 2016 2nd International Conference on Cloud Computing Technologies and Applications (CloudTech). IEEE, Piscataway (2016)
8. Jia, Y., Shelhamer, E., Donahue, J., Karayev, S., Long, J., Girshick, R., Guadarrama, S., Darrell, T.: UC Berkeley Eecs, Caffe: convolutional architecture for fast feature embedding. In: ACM Multimedia (2014)
9. Ke, H., Li, P., Guo, S.: Crowdsourcing on mobile cloud: cost minimization of joint data acquisition and processing. In: 2014 IEEE Conference on Computer Communications Workshops (INFOCOM WKSHPS). IEEE, Piscataway (2014)
10. Krizhevsky, A., Sutskever, I., Hinton Geoffrey, E.: ImageNet classification with deep convolutional neural networks. In: Advances in Neural Information Processing Systems (NIPS) (2012)
11. LeCun, Y., Jackel, L.D., Bottou, L., Cortes, C., Denker, J.S., Drucker, H., Guyon, I., et al.: Learning algorithms for classification: a comparison on handwritten digit recognition. Neural Netw. Stat. Mech. Perspect. **261**, 276 (1995)
12. Poli, R., Kennedy, J., Blackwell, T.: Particle swarm optimization. Swarm Intell. **1**(1), 33–57 (2007)
13. Szegedy, C., Liu, W., Jia, Y., Sermanet, P., Reed, S., Anguelov, D., Erhan, D., Vanhoucke, V., Rabinovich, A.: Going deeper with convolutions. In: IEEE Conference on Computer Vision and Pattern Recognition (2015)
14. Tang, Z., et al.: Energy-efficient transmission scheduling in mobile phones using machine learning and participatory sensing. IEEE Trans. Veh. Technol. **64**(7), 3167–3176 (2015)
15. Yao, D., et al.: Using crowdsourcing to provide QoS for mobile cloud computing. IEEE Trans. Cloud Comput. **7**(2), 344–356 (2015)

Compressed Sensing Channel Estimation for LTE-V

Kelvin Chelli, Ramzi Theodory, and Thorsten Herfet

1 Preliminaries

With the advent of 5G and the convergence of broadcast and broadband technologies, the consequences of high mobility at both the transmitter and receiver along with the methods to compensate the same has become an important consideration in the design and development of modern telecommunication systems. The Release-14 of LTE introduces various improvements to the standard to enable vehicular communication [2].

Vehicular environments are characterized by varying degrees of mobility resulting in a heterogeneous channel. In cases of high mobility, a temporally-varying multipath channel that displays selectivity in both the frequency and time domains whereas, under low mobility a pure frequency selective channel is present. Thus, channel estimation schemes have to work robustly in these heterogeneous channel conditions. Moreover, the computational complexity of these schemes must be relevant for consumer hardware implementation.

In our paper, we develop a scheme for channel estimation that takes into consideration the temporal variations in the channel and that is able to provide good results with normalized Doppler shifts of up to 10%. The *Rake-Matching Pursuit* (RMP) algorithm is a *Compressed Sensing* (CS) scheme that is able to exploit the inherent sparsity of wireless channels and supply a precise estimate of a channel that is doubly selective [4]. On the other hand, a simple scheme like the *Least Squares* (LS) estimator is used in low mobility conditions. Switching between the two schemes is enabled by a simple cognitive framework based on the *Index of Dispersion* that determines the time variation of the channel.

K. Chelli (✉) · R. Theodory · T. Herfet
University in Saarbrücken, Saarbrücken, Germany
e-mail: chelli@nt.uni-saarland.de; herfet@nt.uni-saarland.de

B. Ye et al. (eds.), *2nd International Conference on 5G for Ubiquitous Connectivity*, EAI/Springer Innovations in Communication and Computing,
https://doi.org/10.1007/978-3-030-22316-8_2

This paper is arranged in the following manner. A short summary of the literature survey is provided in Sect. 1.1. The LTE-V system model is introduced in Sect. 1.2. The RMP algorithm for estimating the channel along with the cognitive scheme to adapt estimation is introduced in Sect. 2. The simulation results and the associated computational complexity is consolidated in Sect. 3. Finally, a few concluding remarks are given in Sect. 4.

We assume the subsequent notations in our paper: $\mathbb{C}$ represents a complex number set. While matrices are denoted using upper boldface letters, column vectors are represented by lower boldface letters. The Hermitian transpose is expressed by $(\cdot)^H$. Finally, the ℓ_2-norm of a vector is given by $\|\cdot\|_2$ and the absolute value is represented by $|\cdot|$.

1.1 Literature Survey

Vehicular environments are characterized by heterogeneous channels caused mainly due to changes in mobility. In pure multipath environments, a frequency domain one tap equalization is the conventional choice [15]. However, under high mobility conditions, a time varying multipath channel is present. Estimation and compensation of such channels is a non trivial task and requires more advanced channel estimation schemes [3, 7, 11]. The multidimensional filtering algorithms based on minimum mean square error and the well known *Basis Expansion Model* (BEM) based methods for estimation are shown to work well for such channels. However, these schemes are either too expensive computationally, or provide an estimate that is unable to fully compensate the channel [14]. The *Rake-Matching Pursuit* (RMP) algorithm stems from the theory of *Compressed Sensing* (CS) and is shown to robustly estimate a time varying multipath channel [4–7]. As a consequence of their merits, CS schemes are being studied and developed as a tool for the estimation of doubly selective channels [10, 12, 14]. However, an ideal channel estimator must also work in scenarios of low mobility with a correspondingly lower computational effort. Thus, in such heterogeneous channel conditions an adaptive channel estimation scheme is envisioned and is the goal of this paper.

1.2 The LTE-V System Model

Release-14 of LTE introduces support for V2X to enable robust vehicular communication. The enhancements introduced in this release that are relevant for this paper are the use of SC-FDMA that stands for *Single-Carrier Frequency Division Multiple Access* and an improved pilot pattern. SC-FDMA is closely related to *Orthogonal Frequency-Division Multiple Access* (OFDMA) and is regarded as a DFT-spread OFDMA. It has the same foundations of OFDMA and as such has the same merits with respect to multipath robustness and simple equalization. The only difference is the application of a DFT to the modulated input symbols that has the effect of spreading the symbols over all the subcarriers and thus producing a virtual single carrier structure. This means that in SC-FDMA each subcarrier will carry information about all of the transmitted symbols. Contrastingly in OFDMA, each subcarrier carries information about a single input symbol. As a consequence, a relatively low peak-to-average power ratio compared to OFDMA is achieved. This makes it a good alternative to OFDMA in terms of power consumption by circumventing the need for power amplifiers that are highly linear. Furthermore, since DFT spreading ensures that all subcarriers carry information about all the input signals the robustness to deep narrowband fading is enhanced. On the other hand, IDFT despreading spreads the additive noise power resulting in a phenomenon called noise enhancement at the receiver. Consequently, the performance of the SC-FDMA system is degraded [13]. A typical V2X system is shown in Fig. 1.

One of the most important modifications introduced in V2X is the addition of more reference symbols in order to cope with higher Doppler shifts. Figure 2 shows how the reference symbols are distributed in a subframe. The additional reference symbols aid in channel estimation.

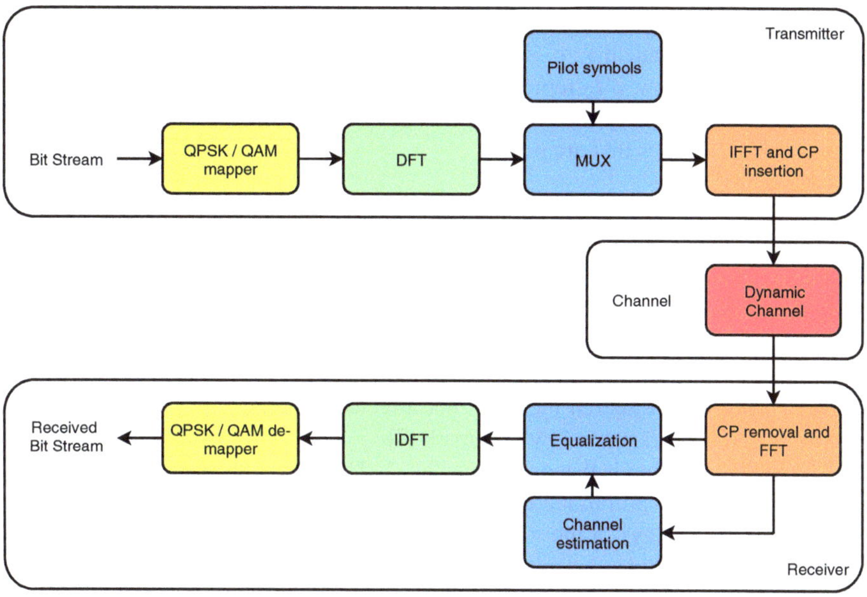

Fig. 1 V2X system model

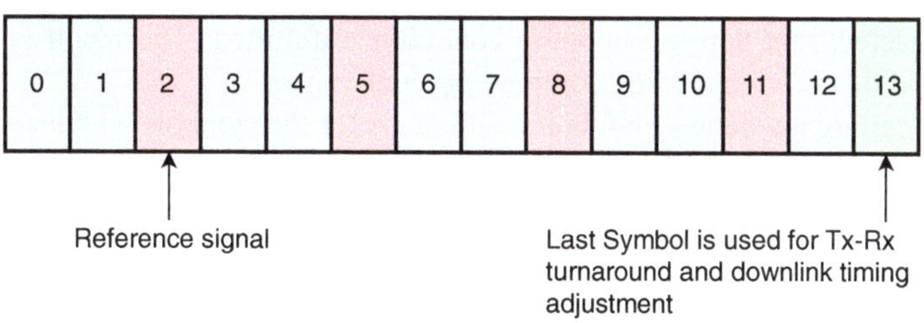

Fig. 2 V2X reference symbols [1]

2 Techniques for Channel Estimation

Learning the properties of the wireless channel by using a set of known symbols called pilots is considered as channel estimation. It plays a vital step in coherent equalization techniques and help in compensating the effects of the underlying channel [9].

2.1 *The Rake-Matching Pursuit Algorithm*

Here we propose the *Rake-Matching Pursuit* (RMP) algorithm to perform estimation of the channel. A dictionary denoted $\mathbf{D}$ serves as a reference database of channel distortions and is a prerequisite for the RMP algorithm. The columns of the dictionary consist of the reference pilot symbols that are corrupted by different amounts of delays. An echo due to multipath results in a delay τ_k at the receiver, which in the frequency domain corresponds to a phase shift dependent on the frequency. Mathematically, it is a multiplication with $e^{-j2\pi f\tau_k}$, where f is the frequency index. Now, if l_p represents the specified symbol containing pilots, every column of the dictionary $d_p \in \mathbb{C}^{M\times 1}$ can be computed by Eq. (1). Here M is the number of subcarriers that are pilots.

$$\boldsymbol{d}_p = \boldsymbol{l}_p \circ e^{-j2\pi f\tau_k} \tag{1}$$

The resulting dictionary $\mathbf{D} \in \mathbb{C}^{M\times K}$ is a collection of the column vectors, where K is the maximum number of delays that have to be searched after an adequate sampling of the underlying delay profile. The proposed delay search metric is amplitude normalized and thus searches only for the phase variations in the signal. This ensures that the search metric is not affected by noise and provides a robust estimate of the delays caused by multipath [7].

The RMP algorithm is listed in Algorithm 1. The rank-1 projections computed in Eq. (3) are robust to amplitude distortions because of the normalization in Eq. (2). Thus the correct delays in the wireless channel are estimated. In the next step, the projections are maximized resulting in an estimate of the first delay tap as in Eq. (4). The complex weighting factor of the delay tap is calculated in Eq. (6). In the next step, the contribution of the previous tap d_{s_p} is subtracted as shown in Eq. (7). In this manner, the algorithm continues until either the maximum allowed iterations are completed or a defined stopping condition is fulfilled. The result is a set of delays and their corresponding complex tap coefficients.

After getting the delays and their coefficients for the symbols which contain the pilots, the channel at these symbol locations can be reconstructed. A Doppler shift introduces a temporal variation of the channel. Estimating the channel at the location

Algorithm 1 Rake-matching pursuit

Initialization

$$r_0 = \frac{y}{|y|} \tag{2}$$

$$b_{0,j} = d_j^H r_0, \quad \text{for } j = 1 \cdots K \tag{3}$$

$$s_1 = \arg\max_{j=1\cdots K} \frac{|b_{0,j}|^2}{\|d_j\|_2} \tag{4}$$

$$i_1 = \{s_1\} \tag{5}$$

$$\hat{x}_1 = \frac{b_{0,s_1}}{\|d_{s_1}\|_2^2} \tag{6}$$

$$b_{1,j} = b_{0,j} - \hat{x}_1 d_j^H d_{s_1}, \quad \text{for } j = 1 \cdots K,\ j \notin i_1 \tag{7}$$

the pth iteration, $p > 1$

$$s_p = \arg\max_{j=1\cdots K, j \notin i_{p-1}} \frac{|b_{p-1,j}|^2}{\|d_j\|_2} \tag{8}$$

$$i_p = \{i_{p-1}, s_p\} \tag{9}$$

$$\hat{x}_p = \frac{b_{p-1,s_p}}{\|d_{s_p}\|_2^2} \tag{10}$$

$$b_{p,j} = b_{p-1,j} - \hat{x}_p d_j^H d_{s_p}, \quad \text{for } j = 1 \cdots K,\ j \notin i_p \tag{11}$$

of the pilot symbols and performing interpolation between them is an implicit method to track the channel and thereby estimate the Doppler shift [7].

2.2 *Cognitive Channel Estimation*

The wireless communication channel is a natural phenomenon and the inherent variation of its characteristics requires a channel estimation scheme that adapts itself accordingly. If the channel is pseudo stationary, a simple low complexity channel estimation scheme is apt, whereas under high mobility, the channel is changing rapidly across time and a more complex channel estimation scheme is required. Cognition is achievable only when the variation of channel characteristics is quantifiable.

The index of dispersion or *Variance to Mean* ratio (VMR) is a normalized measure of dispersion and is proposed to quantify the channel variations. It is defined as $D = \frac{\sigma^2}{\mu}$, where σ^2 is the variance and μ is the mean. The VMR is calculated for a pilot subcarrier across an LTE subframe. To avoid the influence of noise, the VMR is calculated as an average between a set of pilot subcarriers from the received subframe. A single threshold is then used to switch between the RMP and a simple estimation scheme, which in our case is the LS estimator. A VMR value of zero indicates that the data is not dispersed, which means that

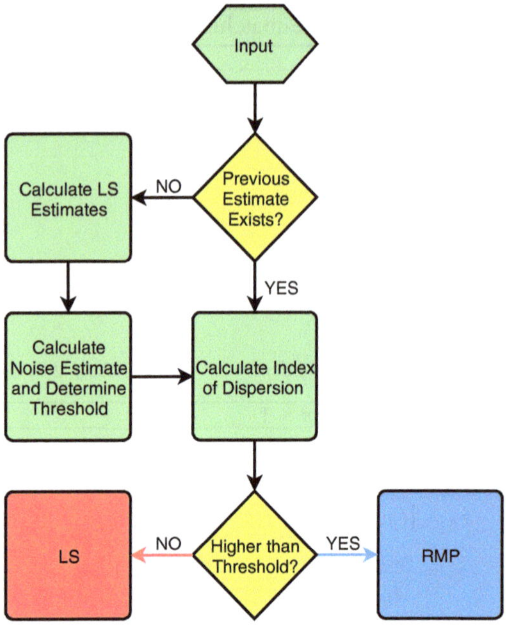

Fig. 3 Decision flowchart

the channel is stationary, a value between 0 and 1 means that the data is under-dispersed, and a value larger than 1 implies that the data is over-dispersed [8]. This means that a threshold value between zero and one can be used in order to switch between the RMP and any arbitrary estimation scheme suited for low mobility. The threshold chooses a channel estimation scheme and thus, a fixed or static threshold is not suitable in the presence of noise. Consequently, a dynamic threshold ξ that is inversely proportional to the noise power estimate N_p is implemented.

The cognitive framework is illustrated by the decision flowchart in Fig. 3. At the beginning of a frame where a previous estimate of the channel is not available, the LS estimates at the pilot locations are calculated, and a noise power estimate N_p is computed that is used to tune and adapt the threshold ξ. Next, the channel variation is quantified by the VMR metric. After thresholding, either the LS or the RMP scheme is used to perform estimation for the subframe. This ensures that at any given subframe, the most optimal channel estimation scheme is employed.

2.3 *Equalization*

The proposed framework for estimation provides the channel transfer function for the physical layer frame. A one-tap equalizer is then applied to compensate the effects of the channel. For an estimate that accurately represents the underlying channel, an adequate compensation of channel distortions is ensured [5].

3 Evaluation

The performance of the proposed framework for estimation is analyzed using the signal to noise ratio versus bit error rate graphs (SNR vs. BER). Standard compliant LTE-V (Release-14), uplink and downlink PHY frames are generated which are then corrupted by the channel. The wireless channel is simulated by the multipath fading model from the Matlab® LTE system toolbox. The *Extended Vehicular A* (EVA) delay profile shown in Table 1 is used along with varying amounts of normalized Doppler shifts up to 10%. The normalized Doppler shift values represent the Doppler shift as a percentage of the subcarrier spacing.

At the receiver, the pilots are extracted from the frame and used for estimation and construct the channel transfer function. The equalizer then attempts to equalize the distortions of the wireless channel.

3.1 Simulation Results

The proposed RMP algorithm along with the cognitive framework is compared to the conventional LS channel estimation under varying mobility conditions. The results for the LTE-Uplink, LTE-Downlink and LTE-V are shown for every channel configuration. Due to the fact that the LTE modes have different physical layer parameters such as the pilot pattern, number of pilot symbols and the choice of the waveform, a direct influence on the effectiveness of estimation methods is expected. Accordingly, the LTE-Downlink mode that specifies an OFDM waveform with a comb-like pilot structure is best suited to track the temporal variations in the channel. The LTE-Uplink and the LTE-V modes both specify SC-FDMA waveforms with a block-type pilot pattern. The LTE-V mode has double the pilot symbols in comparison to the LTE-Uplink and thus performs significantly better under high mobility by being able to track the channel variations more effectively. However, the use of SC-FDMA waveform spreads the channel distortions [13] and thereby

Table 1 EVA profile for delays

Tap delay (μs)	Power (dBm)
0	30.0
0.03	28.5
0.150	28.6
0.310	26.4
0.370	29.4
0.710	20.9
1.090	23.0
1.730	18.0
2.510	13.1

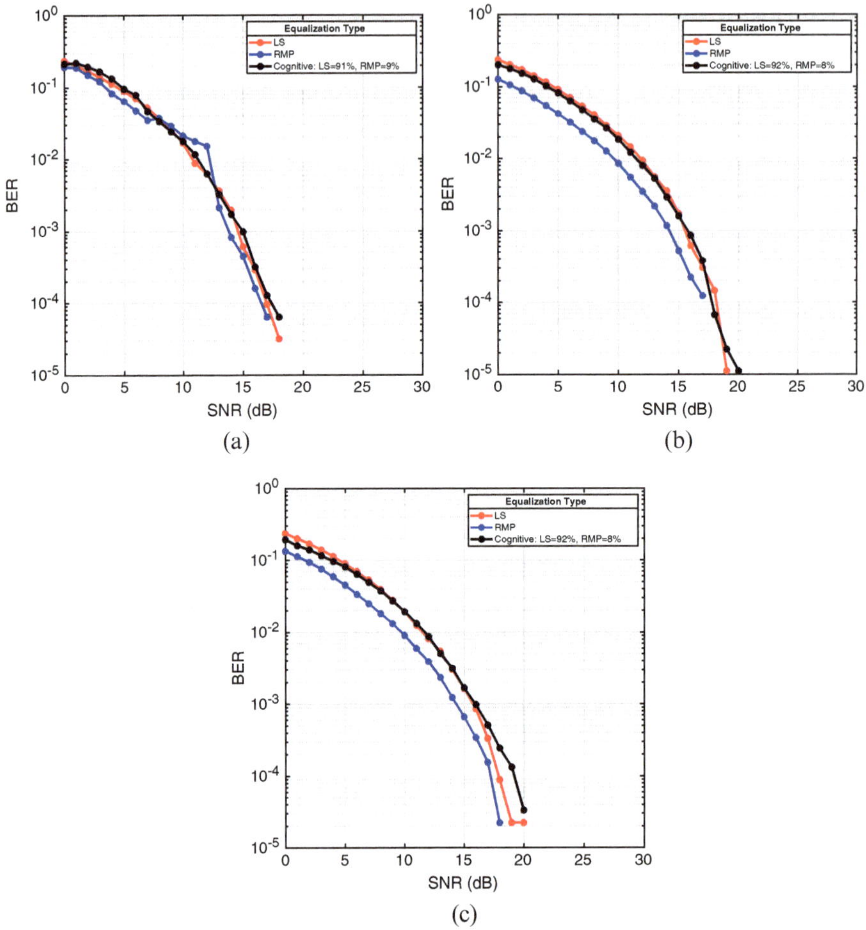

Fig. 4 Comparison for 0% normalized Doppler shift. (**a**) Downlink. (**b**) Uplink. (**c**) V2X pilot pattern

results in a slightly worse performance compared to the LTE-Downlink mode (uses OFDM).

Investigating the performance for a given channel condition, where stationary channel conditions are simulated in Fig. 4, the RMP algorithm as well as the cognitive channel estimation scheme perform equally well. The framework for cognition correctly detects the channel conditions and appropriately chooses the right channel estimation scheme.

Increasing the normalized Doppler to 5% already shows the merits of the RMP algorithm in the estimation of doubly-selective channels when compared to the LS estimator as seen in Fig. 5. A 5% normalized Doppler shift corresponds to 750 Hz and can be considered as a fringe case where, using a single threshold makes it

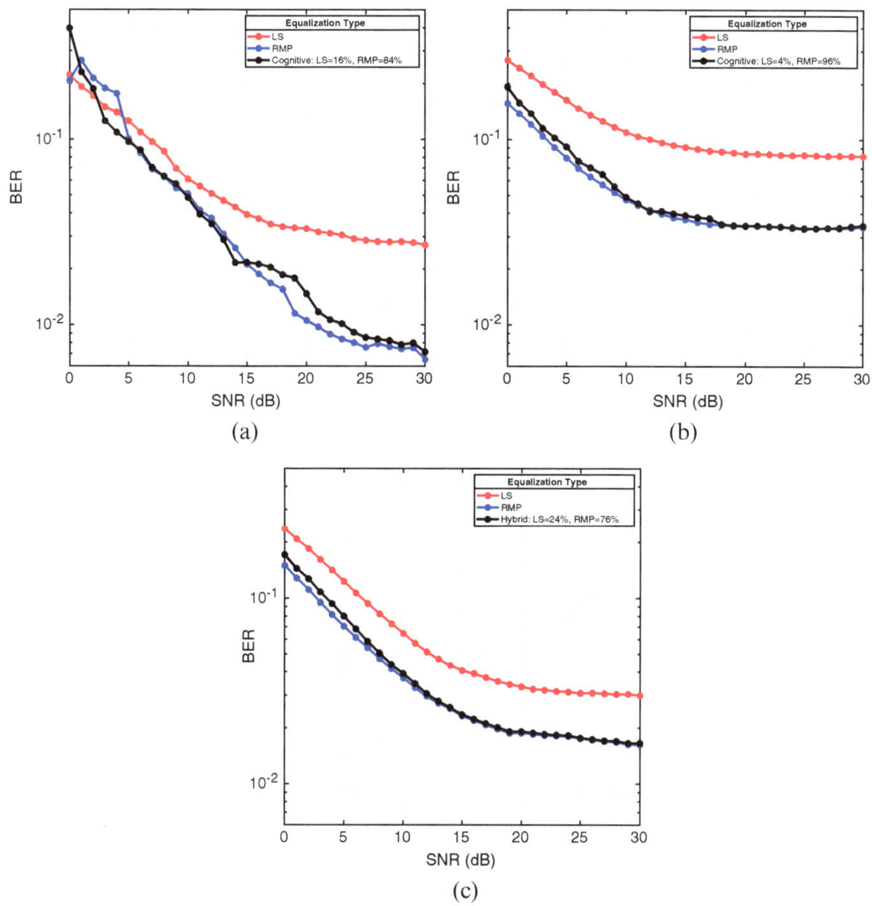

Fig. 5 Comparison for 5% normalized Doppler shift. (**a**) Downlink. (**b**) Uplink. (**c**) V2X pilot pattern

difficult to appoint the correct scheme to perform estimation. Nevertheless, the dynamic VMR metric in the proposed framework accurately quantifies the channel and chooses the RMP algorithm most of the time to estimate the channel.

Finally, for Fig. 6a 1500 Hz Doppler shift that corresponds to a relative velocity of 275 km/h and a normalized Doppler shift of 10% is simulated. The results exhibit a clear gain in performance with the RMP as well as the cognitive framework.

The coherence times for the 5% and 10% normalized Doppler shifts are 666.67 and 333.33 μs respectively, which means that in both cases the channel exhibits temporal variation within a physical layer subframe that has a duration of 1 ms. The results show that the proposed RMP algorithm is capable of precisely estimating varying channel conditions. Moreover, the cognitive framework is able to quantify

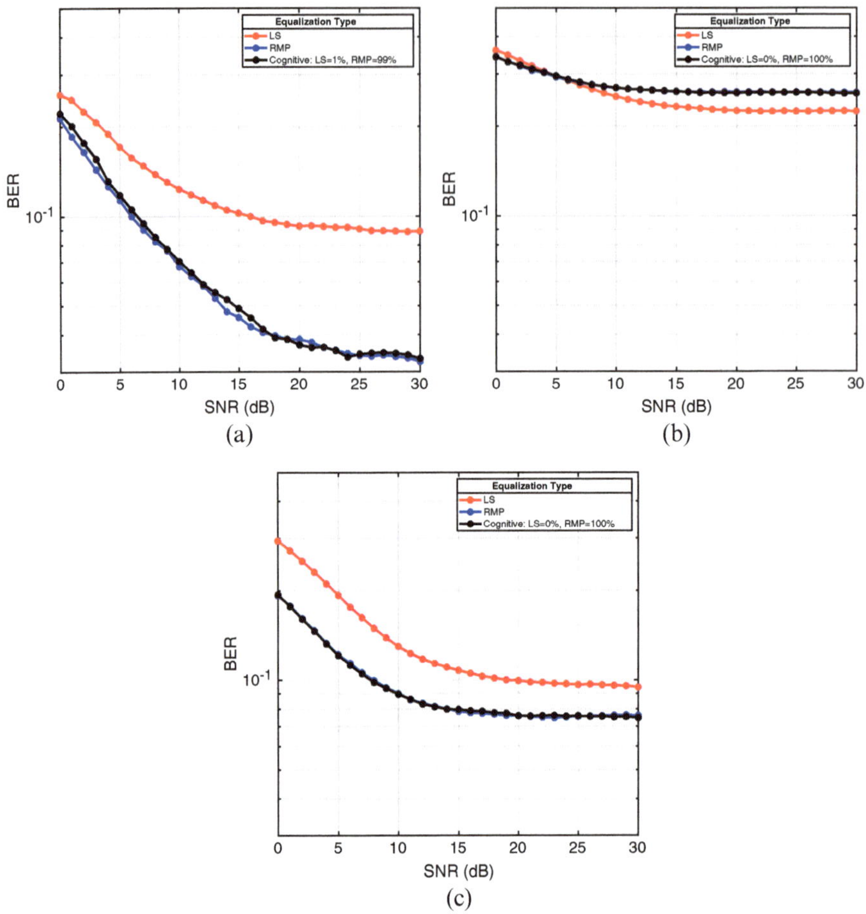

Fig. 6 Comparison for 10% normalized Doppler shift. (**a**) Downlink. (**b**) Uplink. (**c**) V2X pilot pattern

the channel and adapt the channel estimation schemes to keep the complexity relevant while still performing better than the conventional schemes.

3.2 *Complexity*

The computation complexity for LS in each subframe is $\mathcal{O}(PM)$, where M denotes the pilot subcarriers and P are the pilot symbols. However, the computational complexity of the RMP scheme is significantly higher at $\mathcal{O}(PKqM)$, where K represents the maximum number of delays to be searched, and the maximum number of iterations is given by q. In addition to this, linear interpolation is performed to calculate the wireless channel for the data OFDMA/SC-FDMA symbols in the physical layer frame. The cognitive framework further reduces the computational effort by switching between the estimation schemes based on the channel conditions.

4 Conclusion

With the advent of 5G and the inevitable convergence of broadband, broadcast and cellular technologies, a reliable, robust and flexible wireless communication system is envisioned. Consequently a heterogeneous channel with a varying degree of mobility has to be compensated. Here, we suggest a hybrid channel estimation scheme that is able to robustly estimate the wireless channel with an optimal complexity compared to conventional methods. The underlying idea has been to employ channel estimation schemes that are appropriate for a given channel condition. The results not only confirm the benefits of the RMP algorithm under high mobility channel conditions, but also exhibit gains in complexity that are enabled by the cognitive framework. Thus, the proposed schemes are ideally suited for the estimation of a heterogeneous wireless channel with a computational complexity that is viable for implementation on consumer hardware.

References

1. 3GPP: Initial Cellular V2X standard completed (September 2016). http://www.3gpp.org/news-events/3gpp-news/1798-v2x_r14
2. Akyildiz, I.F., Nie, S., Lin, S.C., Chandrasekaran, M.: 5G roadmap. Int. J. Comput. Telecommun. Netw. **106**(C), 17–48 (2016)
3. Berger, C.R., Zhou, S., Preisig, J.C., Willett, P.: Sparse channel estimation for multicarrier underwater acoustic communication: from subspace methods to compressed sensing. IEEE Trans. Signal Process. **58**(3), 1708–1721 (2010). https://doi.org/10.1109/TSP.2009.2038424
4. Chelli, K., Herfet, T.: Doppler shift compensation in vehicular communication systems. In: 2nd IEEE International Conference on Computer and Communications (2016). Available at: http://bit.ly/paper150
5. Chelli, K., Herfet, T.: Estimating doubly-selective channels in DVB-T2. In: 2018 IEEE International Symposium on Broadband Multimedia Systems and Broadcasting (BMSB), June 2018
6. Chelli, K., Sirsi, P., Herfet, T.: Complexity reduction for consumer device compressed sensing channel estimation. In: 2017 IEEE 7th International Conference on Consumer Electronics - Berlin (ICCE-Berlin), September 2017, pp. 189–194. https://doi.org/10.1109/ICCE-Berlin.2017.8210625
7. Chelli, K., Sirsi, P., Herfet, T.: Sparse doubly-selective channels: estimating path parameters unambiguously. In: 2017 European Conference on Networks and Communications (EuCNC): Physical Layer and Fundamentals (PHY) (EuCNC2017 - PHY), Oulu, June 2017
8. Cox, D.R.: The Statistical Analysis of Series of Events. Springer, Dordrecht (1966)
9. Davenport, M.A., Duarte, M.F., Eldar, Y., Kutyniok, G.: Introduction to compressed sensing. In: Compressed Sensing: Theory and Applications. Cambridge University Press ,Cambridge (2011)

10. Ding, W., Yang, F., Pan, C., Dai, L., Song, J.: Compressive sensing based channel estimation for OFDM systems under long delay channels. IEEE Trans. Broadcast. **60**(2), 313–321 (2014)
11. Hrycak, T., Das, S., Matz, G., Feichtinger, H.G.: Low complexity equalization for doubly selective channels modeled by a basis expansion. IEEE Trans. Signal Process. **58**(11), 5706–5719 (2010). https://doi.org/10.1109/TSP.2010.2063426
12. Hu, D., Wang, X., He, L.: A new sparse channel estimation and tracking method for time-varying OFDM systems. IEEE Trans. Veh. Technol. **62**(9), 4648–4653 (2013). https://doi.org/10.1109/TVT.2013.2266282
13. Penttinen, J.T. (ed.): The LTE-Advanced Deployment Handbook: The Planning Guidelines for the Fourth Generation Networks. Wiley, New York (2016)
14. Taubock, G., Hlawatsch, F., Eiwen, D., Rauhut, H.: Compressive estimation of doubly selective channels in multicarrier systems: leakage effects and sparsity-enhancing processing. IEEE J. Sel. Top. Signal Process. **4**(2), 255–271 (2010). https://doi.org/10.1109/JSTSP.2010.2042410
15. Zhao, Z., Cheng, X., Wen, M., Jiao, B., Wang, C.X.: Channel estimation schemes for IEEE 802.11p standard. IEEE Intell. Transport. Syst. Mag. **5**(4), 38–49 (2013). http://dblp.uni-trier.de/db/journals/itsm/itsm5.html#ZhaoCWJW13

Power Allocation Scheme for Non-Orthogonal Multiple Access in Cloud Radio Access Networks

Benben Wen, Tao Liu, Xiangbin Yu, and Fengcheng Xu

1 Introduction

Non-orthogonal multiple access (NOMA) has been considered as a promising candidate multiple access technology for the fifth generation (5G) communication because of the high utilization of resource blocks [1, 2]. Different from the existing orthogonal multiple access (OMA) system, NOMA system can serve multi-users with same frequency and time resource-blocks by making full use of the resources of power domain [3]. The NOMA users can decode signal from the superposed symbols by using successive interference cancellation (SIC) [4].

The cloud radio access networks (C-RANs) is also a key technology in current communication architecture [5, 6]. Compared with the conventional networks, C-RAN has spatially separated remote radio heads (RRHs) distributed in the whole cellular network, which can effectively reduce average access distance and improve spectral efficiency (SE) [7]. However, as the RRH numbers grows, the available resource-blocks are becoming rarer. Hence, it is necessary to implement NOMA technology in C-RAN, i.e., C-RAN-NOMA, to improve the utilization of resource.

There have been many researches for the performance analysis and power allocation in NOMA system. The performance of NOMA system is analyzed in paper [8], where users are randomly deployed in the cellular network. The results show that the SE performance in NOMA is significantly better than that in conventional orthogonal multiple access. The application of NOMA in C-RAN is discussed in [9], and the outage probability is studied and corresponding closed-form approximate expression is derived. The power allocation of NOMA in C-RAN is studied in [10]. However, each user in this system is served by a single RRH,

B. Wen · T. Liu · X. Yu (✉) · F. Xu
College of Electronic and Information Engineering, Nanjing University of Aeronautics and Astronautics, Nanjing, China

B. Ye et al. (eds.), *2nd International Conference on 5G for Ubiquitous Connectivity*, EAI/Springer Innovations in Communication and Computing,
https://doi.org/10.1007/978-3-030-22316-8_3

which may waste some power of other RRHs. To the best of my knowledge, the research on the optimal power allocation scheme design for C-RAN-NOMA is addressed less.

Therefore, the optimal power allocation of a C-RAN-NOMA system is studied in this paper, where two NOMA users served by C-RAN are considered and the RRHs are uniformly distributed in a cellular network. The main contributions of the paper are summarized: (1) By maximizing the sum rate, an optimization model of power allocation for C-RAN-NOMA is proposed under the minimum rate constraints. (2) According to the proposed optimization model, an optimal power allocation scheme is developed, and an efficient method based on the linear programming (LP) algorithm is proposed to obtain the optimal solution of power allocation. (3) Simulation results verify the effectiveness of the developed scheme, and can outperform the conventional equal power scheme.

2 System Model

Consider a downlink C-RAN-NOMA system model similar to the model in [9], as shown in Fig. 1, where N RRHs are uniformly distributed in a disk and serve two NOMA users equipped with single antenna simultaneously in the same time and frequency resource-block. User-1 is located at the center of the cell, i.e., center-

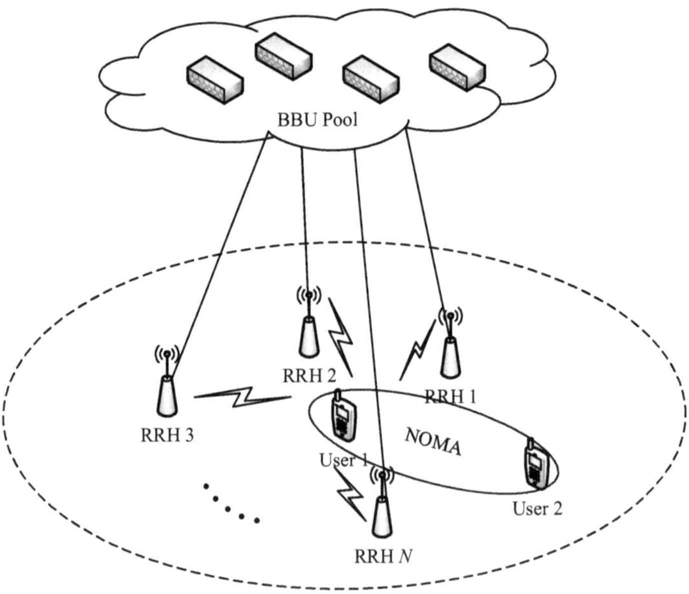

Fig. 1 A simplified C-RAN-NOMA system model

user. User-2 is located at the edge region of the cell, i.e., edge-user. In other words, User-1 has better channel condition than User-2 for most of the RRHs.

In the proposed system model, the superposed signal transmitted by RRH-i is given by

$$x_i = \sqrt{a_1^i P} s_1 + \sqrt{a_2^i P} s_2, \tag{1}$$

where a_j^i denotes the power allocation coefficient for User-j on RRH-i, i.e., $a_1^i + a_2^i = 1$, and $a_1^i \leq a_2^i$ should be ensured in NOMA system [8], P is the transmit power of each RRH, s_1, s_2 denote the desired signal of User-1 and User-2, respectively.

Both the small-scale and large-scale fading are considered in this paper. Therefore, the channel between RRH-i and User-j can be modeled as

$$h_j^i = \frac{\tilde{h}_j^i}{\sqrt{1 + \left(d_j^i\right)^\alpha}}, \tag{2}$$

where d_j^i denotes the distance between RRH-i and User-j, α is the path loss exponent, $\tilde{h}_j^i$ represents the Rayleigh fading coefficient between RRH-i and User-j, which is modeled as an independent and identically distributed complex Gaussian random variable, i.e., $\tilde{h}_j^i \sim \mathcal{CN}(0, 1)$. Let $g_j^i = \left|h_j^i\right|^2$ denote the channel gain between RRH-i and User-j.

Therefore, the received signals at User-1 and User-2 are given by

$$\begin{cases} y_1 = \left(\sum\limits_{i=1}^{N} h_1^i \sqrt{a_1^i P}\right) s_1 + \left(\sum\limits_{i=1}^{N} h_1^i \sqrt{a_2^i P}\right) s_2 + n_1, \\ y_2 = \left(\sum\limits_{i=1}^{N} h_2^i \sqrt{a_1^i P}\right) s_1 + \left(\sum\limits_{i=1}^{N} h_2^i \sqrt{a_2^i P}\right) s_2 + n_2, \end{cases} \tag{3}$$

where n_i represents the complex Gaussian noise with zero-mean and variance σ^2, i.e., $n_i \sim \mathcal{CN}\left(0, \sigma^2\right)$.

According to the principle of NOMA, User-1 will first decode the signal of User-2 from the superposed signal and then decode s_1 without the interference of User-2, while User-2 decodes the desired signal s_2 from the superposed signal directly by treating the signal of User-1 as noise [11]. Therefore, the sum rate of User-1 and User-2 can be expressed as

$$R_S = R_1 + R_2, \tag{4}$$

where R_1 and R_2 denote the achievable rates of User-1 and User-2, respectively, which are given by

$$\begin{cases} R_1 = \log_2\left(1 + \frac{P\sum_{i=1}^{N} a_1^i g_1^i}{\sigma^2}\right), \\ R_2 = \log_2\left(1 + \frac{P\sum_{i=1}^{N} a_2^i g_2^i}{P\sum_{i=1}^{N} a_1^i g_2^i + \sigma^2}\right). \end{cases} \tag{5}$$

3 Power Allocation for Sum Rate Maximization

Based on the system model proposed in Sect. 2, the optimization problem for the sum rate maximization can be formulated as

$$\begin{aligned} \max_{\{a_1^i, a_2^i\}} \quad & R_S = \log_2\left(1 + \frac{P\sum_{i=1}^{N} a_1^i g_1^i}{\sigma^2}\right) + \log_2\left(1 + \frac{P\sum_{i=1}^{N} a_2^i g_2^i}{P\sum_{i=1}^{N} a_1^i g_2^i + \sigma^2}\right) \\ \text{s.t.} \quad & a_1^i + a_2^i = 1, \quad 1 \le j \le N, \\ & a_1^i \le a_2^i, \quad 1 \le j \le N, \\ & R_1 \ge R_1^{\min}, \\ & R_2 \ge R_2^{\min}, \end{aligned} \tag{6}$$

where $R_i^{\min}$ denotes the minimum rate constraint of User-i.

The optimization problem (6) can be transformed into the following problem (7):

$$\begin{aligned} \max_{\mathbf{a}_1} \quad & R_S(\mathbf{a}_1) = \log_2\left(\mathbf{g}_1^{\mathrm{T}}\mathbf{a}_1 P + \sigma^2\right) - \log_2\left(\mathbf{g}_2^{\mathrm{T}}\mathbf{a}_1 P + \sigma^2\right) + \log_2\left(1 + \frac{P\sum_{i=1}^{N} g_2^i}{\sigma^2}\right) \\ & = f_{cave}(\mathbf{a}_1) + f_{vex}(\mathbf{a}_1) + \log_2\left(1 + P\textstyle\sum_{i=1}^{N} g_2^i/\sigma^2\right) \\ \text{s.t.} \quad & C_1 : \mathbf{0}_{N\times 1} \preccurlyeq \mathbf{a}_1 \preccurlyeq \tfrac{1}{2}\mathbf{1}_{N\times 1}, \\ & C_2 : \mathbf{g}_1^{\mathrm{T}}\mathbf{a}_1 \ge \left(2^{R_1^{\min}} - 1\right)\sigma^2/P, \\ & C_3 : \mathbf{g}_2^{\mathrm{T}}\mathbf{a}_1 \le \frac{\mathbf{g}_2^{\mathrm{T}}\mathbf{1}_{N\times 1}P + \sigma^2}{2^{R_2^{\min}} P} - \frac{\sigma^2}{P}, \end{aligned} \tag{7}$$

where $\mathbf{a}_1 = \left[a_1^1, \ldots, a_1^N\right]^{\mathrm{T}}$, $\mathbf{g}_i = \left[g_i^1, \ldots, g_i^N\right]^{\mathrm{T}}$, $\mathbf{0}_{N\times 1} = \left[\underbrace{0, \ldots, 0}_{N}\right]^{\mathrm{T}}$, $\mathbf{1}_{N\times 1} = \left[\underbrace{1, \ldots, 1}_{N}\right]^{\mathrm{T}}$ and

$$\begin{cases} f_{cave}(\mathbf{a}_1) = \log_2\left(\mathbf{g}_1^{\mathrm{T}}\mathbf{a}_1 P + \sigma^2\right), \\ f_{vex}(\mathbf{a}_1) = -\log_2\left(\mathbf{g}_2^{\mathrm{T}}\mathbf{a}_1 P + \sigma^2\right). \end{cases} \tag{8}$$

Although the third item of the objective function of (7) is constant during the transmission, $R_S(\mathbf{a}_1)$ is the difference of convex functions, which is inconvenient to solve directly. The exhaustive search algorithm can be adopted here with much higher complexity. To reduce the computational complexity, a more appropriate algorithm should be proposed. Firstly, we transform the objective function of (7) into

$$\begin{aligned} R_S(\mathbf{a}_1) &= \log_2\left(\frac{\mathbf{g}_1^{\mathrm{T}}\mathbf{a}_1 P+\sigma^2}{\mathbf{g}_2^{\mathrm{T}}\mathbf{a}_1 P+\sigma^2}\right)+\log_2\left(1+\frac{P\sum_{i=1}^{N} g_2^i}{\sigma^2}\right) \\ &= \log_2\left[f_{lin}(\mathbf{a}_1)\right]+\log_2\left(1+P\sum_{i=1}^{N} g_2^i/\sigma^2\right), \end{aligned} \tag{9}$$

where $f_{lin}(\mathbf{a}_1) = \left(\mathbf{g}_1^{\mathrm{T}}\mathbf{a}_1 P+\sigma^2\right)/\left(\mathbf{g}_2^{\mathrm{T}}\mathbf{a}_1 P+\sigma^2\right)$. The maximization of (9) is equivalent to the maximization of $f_{lin}(\mathbf{a}_1)$. Moreover, $f_{lin}(\mathbf{a}_1)$ can be converted into

$$f_{lin}(\mathbf{a}_1) = \frac{\mathbf{g}_1^{\mathrm{T}}\mathbf{a}_1 P+\sigma^2}{\mathbf{g}_2^{\mathrm{T}}\mathbf{a}_1 P+\sigma^2} - 1 + 1 = \frac{\left(\mathbf{g}_1^{\mathrm{T}}-\mathbf{g}_2^{\mathrm{T}}\right)\mathbf{a}_1 P}{\mathbf{g}_2^{\mathrm{T}}\mathbf{a}_1 P+\sigma^2} + 1. \tag{10}$$

Let $\mathbf{b}_1 = \frac{\mathbf{a}_1}{\mathbf{g}_2^{\mathrm{T}}\mathbf{a}_1 P+\sigma^2}$, $\mathbf{c} = \left(\mathbf{g}_1^{\mathrm{T}}-\mathbf{g}_2^{\mathrm{T}}\right)P$, $\tilde{f}_{lin}(\mathbf{b}_1) = \mathbf{c}\mathbf{b}_1 + 1$, then we can obtain $\tilde{f}_{lin}(\mathbf{b}_1) = f_{lin}(\mathbf{a}_1)$. For a given $\mathbf{b}_1$, a unique $\mathbf{a}_1$ can be derived as

$$\mathbf{a}_1 = \sigma^2\left(\mathbf{I}_N - \mathbf{b}_1\mathbf{g}_2^{\mathrm{T}}P\right)^{-1}\mathbf{b}_1. \tag{11}$$

Based on the Sherman-Morrison formula, (11) can be simplified as

$$\mathbf{a}_1 = \sigma^2\left(\mathbf{I}_N + \frac{\mathbf{b}_1\mathbf{g}_2^{\mathrm{T}}P}{1-\mathbf{g}_2^{\mathrm{T}}\mathbf{b}_1 P}\right)\mathbf{b}_1. \tag{12}$$

Thus, the problem (7) is equivalent to

$$\begin{aligned} \max_{\mathbf{b}_1}\ & \tilde{f}_{lin}(\mathbf{b}_1) \\ \text{s.t.}\ & C_4 : \mathbf{b}_1 \succcurlyeq \mathbf{0}_{N\times 1}, \\ & C_5 : \mathbf{g}_2^{\mathrm{T}}\mathbf{b}_1 P \le 1, \\ & C_6 : \left(\sigma^2 + \tfrac{1}{2}\mathbf{1}_{N\times 1}\mathbf{g}_2^{\mathrm{T}}P\right)\mathbf{b}_1 \preccurlyeq \tfrac{1}{2}\mathbf{1}_{N\times 1}, \\ & C_7 : \left[\sigma^2\mathbf{g}_1^{\mathrm{T}} + \left(2^{R_1^{\min}}-1\right)\sigma^2\mathbf{g}_2^{\mathrm{T}}\right]\mathbf{b}_1 \ge \left(1-2^{R_1^{\min}}\right)\sigma^2/P, \\ & C_8 : \left(\mathbf{g}_2^{\mathrm{T}}\mathbf{1}_{N\times 1}P+\sigma^2\right)\mathbf{g}_2^{\mathrm{T}}\mathbf{b}_1 \le \mathbf{g}_2^{\mathrm{T}}\mathbf{1}_{N\times 1}P + \left(1-2^{R_2^{\min}}\right)\sigma^2/P, \end{aligned} \tag{13}$$

where the constraints C_4, C_5, C_6 are derived from the constraint C_1, and the constraints C_7, C_8 are derived from constraints C_2, C_3, respectively.

Problem (13) is a standard LP problem and the procedure for solving LP problem is already quite mature, such as simplex algorithm and Interior-point method and so on. For convenience, the "linprog" tool in Matlab is utilized in this paper. Based on the obtained optimal solution $\mathbf{b}_1^*$ of (13), the optimal solution of (7) can be derived by (12).

4 Numerical Results

In this section, the performance of the proposed power allocation scheme for the sum rate maximization is evaluated through computer simulation. Without loss of generality, we consider a scenario that $N = 3/5/7$ RRHs are distributed in the cell. The radius of the cell is set to 500 m [10]. The path loss exponent is $\alpha = 4$ and the power of noise is $\sigma^2 = -104$dBm. The minimum rate constraints of User-1 and User-2 are, respectively, $R_1^{\min} = R_2^{\min} = 3$bit/s/Hz. Each simulation figure is carried out based on 10^5 channel realizations.

Figure 2 gives the performances of C-RAN-NOMA with the proposed power allocation scheme based on the LP algorithm, exhaustive search algorithm and the equal power allocation scheme with $N = 3$. It can be observed that the scheme with equal power allocation has lowest sum rate, while our proposed scheme can even achieve better performance than the inefficient exhaustive algorithm. This is because the equal power scheme does not consider the difference of channel condition

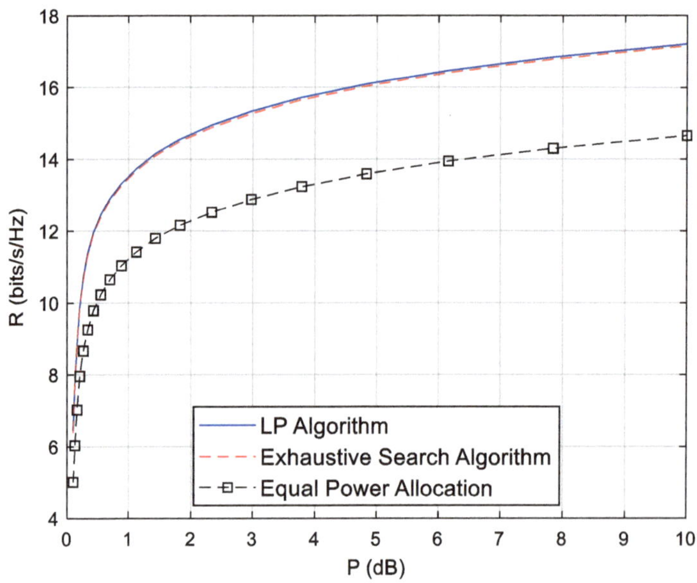

Fig. 2 Comparison of sum rate with different power allocation schemes

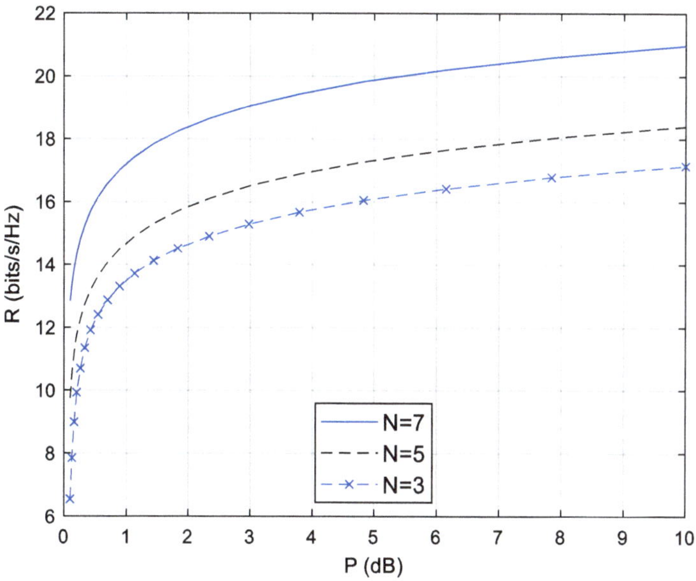

Fig. 3 Comparison of sum rate with different number of RRHs

among different RRHs, and the channel state information is not fully utilized, which results in worst performance. Moreover, our scheme has the same rate as the exhaustive search algorithm, but the latter has much higher complexity because it needs to perform multi-dimensional search for achieving superior performance, especially for large number of RRH, N. Whereas for our scheme, it is obtained based on the maximization of sum rate and can adapt to the change of channel state information. Thus, it exhibits superior performance over the equal power allocation scheme, which implies the validity of the proposed scheme.

Figure 3 shows the comparison of sum rate with different number of RRHs. It can be found that as the N increases the performance of sum rate has notable improvement. It is because the raise of RRHs will introduce more diversity gain, and higher rate can be attained. Figure 4 compares the rate performance of User-1 and User-2 with the proposed scheme. Firstly, we can find that the rates of User-1 and User-2 can both satisfy the minimum rate constraint which is set to 3 bit/s/Hz in the simulation, and the rates become higher with the increase of transmit power P. Besides, User-1 has superior rate performance over User-2 because of better channel condition and SIC procedure. What is more, the increment of N brings the improvement of rate for either User-1 or User-2. The reason is already shown in the analysis of Fig. 3.

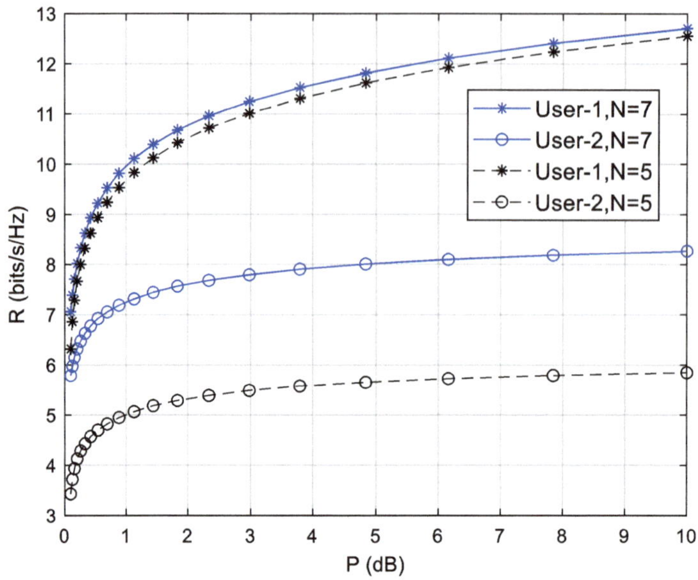

Fig. 4 Comparison of the rates of User-1 and User-2

5 Conclusion

In this paper, the optimal power allocation scheme for maximizing the sum rate in C-RAN-NOMA with two users is developed, and with this scheme, a LP algorithm is proposed to achieve the optimal performance. The simulation results validate the effectiveness of the proposed scheme and algorithm. In the future, we will address the power allocation and pairing scheme designs for the C-RAN-NOMA system with more users for catering to the demand of 5G communications.

Acknowledgements This work is supported in part by National Natural Science Foundation of China (61601220), Natural Science Foundation of Jiangsu Province in China (BK20181289), and National Key Scientific Instrument and Equipment Development Project (Grant No. 2013YQ200607).

References

1. Islam, S.M.R., Avazov, N., Dobre, O.A., Kwak, K.S.: Power-domain non-orthogonal multiple access (NOMA) in 5G systems: potentials and challenges. IEEE Commun. Surv. Tutorials. **19**, 721–742 (2017)
2. Benjebbour, A., Li, A., Saito, K., Saito, Y., Kishiyama, Y., Nakamura, T.: NOMA: from concept to standardization. In: Standards for Communications and Networking, pp. 18–23 (2016)

3. Ding, Z., Peng, M., Poor, H.V.: Cooperative non-orthogonal multiple access in 5g systems. IEEE Commun. Lett. **19**, 1462–1465 (2015)
4. Saito, Y., Kishiyama, Y., Benjebbour, A., Nakamura, T.: Non-orthogonal multiple access (NOMA) for cellular future radio access. In: Vehicular Technology Conference, pp. 1–5 (2017)
5. Checko, A., Christiansen, H.L., Yan, Y., Scolari, L.: Cloud RAN for mobile networks—a technology overview. IEEE Commun. Surv. Tutorials. **17**, 405–426 (2015)
6. Rost, P., Bernardos, C.J., Domenico, A.D., Girolamo, M.D.: Cloud technologies for flexible 5G radio access networks. IEEE Commun. Mag. **52**, 68–76 (2014)
7. Peng, M., Zhang, K., Jiang, J., Wang, J., Wang, W.: Energy-efficient resource assignment and power allocation in heterogeneous cloud radio access networks. IEEE Trans. Veh. Technol. **64**, 5275–5287 (2015)
8. Ding, Z., Yang, Z., Fan, P., Poor, H.V.: On the performance of non-orthogonal multiple access in 5G systems with randomly deployed users. IEEE Signal Process. Lett. **21**, 1501–1505 (2014)
9. Gu, X., Ji, X., Ding, Z., Wu, W., Peng, M.: Outage probability analysis of non-orthogonal multiple access in cloud radio access networks. IEEE Commun. Lett. **22**, 149–152 (2018)
10. Singh, R., Zhu, H., Wang, J.: Performance of non-orthogonal multiple access (NOMA) in a C-RAN system. In: IEEE Vehicular Technology Conference, pp. 1–5 (2017)
11. Zhang, Y., Wang, H.M., Zheng, T.X., Yang, Q.: Energy-efficient transmission design in non-orthogonal multiple access. IEEE Trans. Veh. Technol. **66**, 2852–2857 (2017)

Energy Efficient Optimization Scheme for Uplink Distributed Antenna System with D2D Communication

Guangying Wang, Tao Teng, Xiangbin Yu, and Qiuming Zhu

1 Introduction

In recent years, the number of communication applications is drastically increasing with a higher energy consumption. To satisfy the rising demands of data rate, the fifth generation (5G) mobile communication has been put forward. Energy efficiency (EE) is one of the key performance indicators in 5G networks [1, 2]. Therefore, it is essential to reconsider the existing cellular system and propose new schemes to meet the needs of 5G mobile communication. In contrast with the traditional centralized antenna system (CAS), the distributed antenna system (DAS) has been proved as a promising architecture [3, 4], which can obtain higher EE. A vital technology of 5G is D2D communication, which allows direct communication between two proximate devices [5]. Both of them are very flexible techniques to improve the EE, can reduce the delay and decrease the energy consumption in future communication system.

There are many literatures aiming at the EE in D2D communication. In [6], the authors optimize the total throughput in D2D communication underlaying cellular system. To fulfill the optimization problem, the resource sharing scheme is proposed. Taking the power control into consideration, [7] develops an analytical approach to optimize the EE, which can mitigate the interference between the D2D user (DU) and the cellular user (CU). In [8], the optimal power control scheme in cellular network with D2D communication is proposed to maximize the energy efficiency of DU. The authors in [9] investigate the energy-efficient resource allocation with D2D communication in the CAS. In order to solve the optimal power allocation problem and improve the EE, the authors use fractional programming

G. Wang · T. Teng · X. Yu (✉) · Q. Zhu
College of Electronic and Information Engineering, Nanjing University of Aeronautics and Astronautics, Nanjing, China

B. Ye et al. (eds.), *2nd International Conference on 5G for Ubiquitous Connectivity*, EAI/Springer Innovations in Communication and Computing,
https://doi.org/10.1007/978-3-030-22316-8_4

method. The system performance of CAS with D2D communication and downlink resources allocation are respectively studied in [10, 11]. However, there are few works addressing uplink resources optimization in DAS with D2D communication.

In this paper, we study an energy-efficient optimization for D2D communication in distributed antenna system, and the corresponding optimal power allocation is developed. A DAS model including D2D communication is considered, where a D2D pair reuses the uplink channel resources of DAS. We formulate an optimization problem based on the maximum EE, which considers the maximum power constraint and the minimum rate constraints of CU and DU. Considering the pseudo-concave of objective function in optimization problem, we propose an effective search algorithm with gradient descent method and Armijo method to maximize the EE. Simulation results are shown to the effectiveness of the proposed scheme and algorithm.

The rest of this paper is as follows. The DAS model with D2D communication is established in Sect. 2. In Sect. 3, we formulate the optimization problems for maximizing EE subject to the minimum CU and D2D rates. An optimal algorithm with gradient descent method and Armijo method is proposed. In Sect. 4, simulation results are presented to demonstrate the effectiveness of the proposed algorithm. Finally, we conclude the paper in Sect. 5.

2 System Model

We consider a single-cell uplink DAS with D2D communication, where one CU and a D2D pair (DU1 and DU2) share the spectrum resources as illustrated in Fig. 1. RA_i, $i = 1, \ldots, N$ is the i-th remote antenna distributed randomly in the cell, and both the CU and DU are equipped with single antenna. For simplicity, DU1 and DU2 are denoted as D2D-T and D2D-R, respectively.

According to the system model, the achievable rate of CU is expressed as

$$R_c = \log_2\left(1 + \frac{p\sum_{i=1}^{N} g_{i,c}}{q\sum_{i=1}^{N} g_{i,d} + \sigma_n^2}\right) \tag{1}$$

and the achievable rate of D2D is written as

$$R_d = \log_2\left(1 + \frac{q g_{dd}}{p g_{cd} + \sigma_n^2}\right) \tag{2}$$

where p denotes the transmit power of CU, q denotes the transmit power of D2D-T. $g_{i,c}$ is channel power gain from CU to RA_i, $g_{i,d}$ is the channel power gain from D2D-T to RA_i, g_{dd} is the channel power gain from D2D-T to D2D-R, g_{cd} is the channel power gain from CU to D2D-R, and σ_n^2 is the noise power. The system

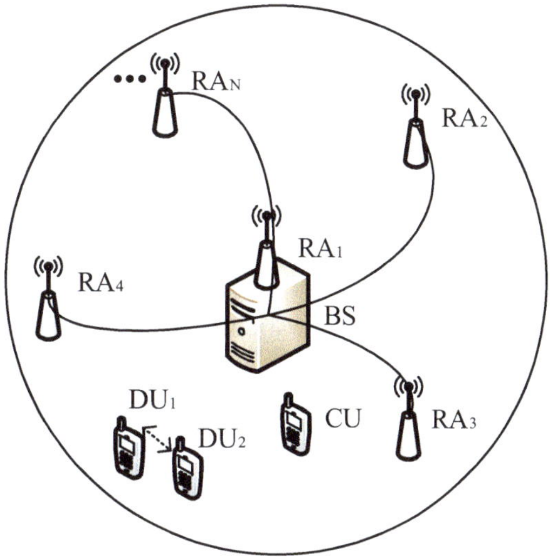

Fig. 1 System model of DAS with D2D

has maximum total power constraint $P_{\max}$, and thus $p + q \le P_{\max}$. Without loss of generality, we normalize the total system bandwidth into unit.

We consider the composite fading channel, which includes path loss and Rayleigh fading. The channel power gain $g_{i,c}$ is written as

$$g_{i,c} = d_{i,c}^{-\beta}\left|h_{i,c}\right|^2 \tag{3}$$

where $d_{i,c}$ is the distance from CU to RA_i. β is the path loss factor. $h_{i,c}$ denotes the small-scale fading coefficient between CU to RA_i, which can be modeled as complex Gaussian random variables with zero mean and unit variance.

Similarly, $g_{i,d} = d_{i,d}^{-\beta}\left|h_{i,d}\right|^2$, $g_{cd} = d_{cd}^{-\beta}|h_{cd}|^2$ and $g_{dd} = d_{dd}^{-\beta}|h_{dd}|^2$ represent the channel power gain between D2D-T and RA_i, CU and D2D-R, D2D-T and D2D-R, respectively. Where $d_{i,d}$, d_{cd}, d_{dd} and $h_{i,d}$, h_{cd}, h_{dd} are the distances and the small-scale fading coefficients between D2D-T and RA_i, CU and D2D-R, D2D-T and D2D-R, respectively.

3 Power Allocation for EE Maximization

According to the system model, an optimization problem for maximizing EE is established. The maximum total power constraint and the minimum rate constraints of the CU and DU are considered. The optimization problem is expressed as

$$\begin{aligned} \max_{p,q} \quad & \eta_{EE} = \tfrac{1}{p+q+P_c}\left(R_c + R_d\right) \\ s.t. \quad & R_c \geq R_{\min,c},\, R_d \geq R_{\min,d}, \\ & 0 \leq p + q \leq P_{\max}, \\ & p \geq 0, \quad q \geq 0 \end{aligned} \tag{4}$$

where η_{EE} is the EE. P_c is the fixed circuit power consumption of system. $R_{\min,\,c}$ denotes the minimum rate of CU and $R_{\min,\,d}$ denotes the minimum rate of DU.

Let $p = \alpha P$, $q = (1 - \alpha)P$, and $0 \leq \alpha \leq 1$, then $P \leq P_{\max}$. Using the transformation of the variables above, we can obtain:

$$R_c = \log_2\left(1 + \frac{m_1 \alpha P}{m_2\left(1-\alpha\right)P + 1}\right) \tag{5}$$

$$R_d = \log_2\left(1 + \frac{m_3\left(1-\alpha\right)P}{m_4 \alpha P + 1}\right) \tag{6}$$

where $m_1 = \frac{\sum_{i=1}^{N} g_{i,c}}{\sigma_n^2}$, $m_2 = \frac{\sum_{i=1}^{N} g_{i,d}}{\sigma_n^2}$, $m_3 = \frac{g_{dd}}{\sigma_n^2}$, $m_4 = \frac{g_{cd}}{\sigma_n^2}$.

With (5) and (6), the problem (4) can be changed into

$$\max_{P} \quad \eta_{EE} = \frac{1}{P + P_c}\left(R_c + R_d\right) \tag{7}$$

$$\begin{aligned} s.t. \quad & R_c \geq R_{\min,c},\, R_d \geq R_{\min,d}, \\ & 0 \leq P \leq P_{\max} \end{aligned} \tag{8}$$

Corollary 1 *η_{EE} is pseudo-concave in P.*

Proof Let $M_1 = m_1\alpha + m_2(1 - \alpha)$, $M_2 = m_2(1 - \alpha)$, $M_3 = m_3(1 - \alpha) + m_4\alpha$, $M_4 = m_4\alpha$, then the sum of rate of CU and DU, R in (7) is given by

$$R = R_c + R_d = \log_2\left(1 + M_1 P\right) - \log_2\left(1 + M_2 P\right) + \log_2\left(1 + M_3 P\right) - \log_2\left(1 + M_4 P\right)$$

For the given α, we can calculate the first derivative and the second derivative of R with respect to P as follows:

$$\frac{\partial R}{\partial P} = \frac{M_1}{1 + M_1 P} - \frac{M_2}{1 + M_2 P} + \frac{M_3}{1 + M_3 P} - \frac{M_4}{1 + M_4 P} \tag{9}$$

$$\frac{\partial^2 R}{\partial P^2} = \frac{\left(M_1 + M_2 + 2M_1M_2P\right)\left(M_2 - M_1\right)}{(1 + M_1P)^2(1 + M_2P)^2} + \frac{\left(M_3 + M_4 + 2M_3M_4P\right)\left(M_4 - M_3\right)}{(1 + M_3P)^2(1 + M_4P)^2} \tag{10}$$

Since $(M_1 + M_2 + 2M_1M_2P) > 0, (M_3 + M_4 + 2M_3M_4P) > 0, M_2 < M_1, M_4 < M_3$ we can get $\frac{\partial^2 R}{\partial P^2} < 0$. Therefore, R is strictly concave in P. In addition, the denominator is linear in equation (7). Thus, η_{EE} is a pseudo-concave function of P.

Corollary 2 *For a given P, the upper and lower bounds of* α *are* $[\alpha_1, \alpha_2] \subset [0, 1]$.

Proof Considering the constraints of minimum rate $R_c \geq R_{\min,\, c}, R_d \geq R_{\min,\, d}$, with (5) and (6), it is easily proved that $\alpha_1 \leq \alpha \leq \alpha_2$, where $\alpha_1 = \frac{r_1 + m_2 r_1 P}{m_1 P + m_2 r_1 P}$, $\alpha_2 = \frac{m_3 P - r_2}{m_3 P + m_4 r_2 P}$, $r_1 = 2^{R_{c,\min}} - 1$ and $r_2 = 2^{R_{d,\min}} - 1$. Hence, we have $\alpha \in [\alpha_1, \alpha_2]$.

Utilizing $\alpha_1 \leq \alpha_2$, we can obtain that $P_{\min} \leq P \leq P_{\max}$, where $P_{\min} > 0$ and $P_{\min} = \frac{m_1 r_2 + m_2 r_1 r_2 + m_4 r_1 r_2 + m_3 r_1}{m_1 m_3 - m_2 m_4 r_1 r_2}$. For the given $P \in [P_{\min}, P_{\max}]$, we have:

$$\eta_{EE}(\alpha, P) = \frac{1}{P + P_c}\left(\log_2\left(1 + \frac{m_1 \alpha P}{m_2(1-\alpha)P+1}\right) + \log_2\left(1 + \frac{m_3(1-\alpha)P}{m_4 \alpha P + 1}\right)\right) \tag{11}$$

With Corollary 1 and Corollary 2, the optimization problem can be solved effectively. For this reason, we present an efficient algorithm based on gradient descent method and Armijo method. The procedure is summarized in Algorithm 1.

Algorithm 1 The Optimization Algorithm

1: For given $P \in [P_{\min}, P_{\max}]$, set the step ε_P and the best EE η_{EE}^P under P

2: For $\alpha_l \in [\alpha_1, \alpha_2]$, set the step ε_α and the best solution P_l^*

(2.1) Set initial point $P^{(0)}$, the initial step $\varepsilon^{(0)}$, termination error ξ and the iteration index $k = 0$

(2.2) Calculate the negative gradient $s^{(k)} = -\nabla \eta_{EE}(\alpha_l, P^{(k)})$

(2.3) Calculate the best step $\varepsilon^{(k)}$ by Armijo method, $P^{(k+1)} = P^{(k)} - \varepsilon^{(k)} s^{(k)}$

(2.4) **If** $|P^{(k+1)} - P^{(k)}| < \xi$, go to (2.5)
else $k = k + 1$, $P^{(k)} = P^{(k+1)}$, go to (2.2)

(2.5) Output the optimal solution P_l

(2.6) **If** $P_l > P$ **then** $P_l^* = P$
else if $P_l \leq P$ **then**
if $R_c(\alpha_l, P_l) < R_{\min,\, c}$ & $R_d(\alpha_l, P_l) \geq R_{\min,\, d}$, **then** $P_l^* = \frac{r_1}{m_1\alpha_l - m_2 r_1(1-\alpha_l)}$
if $R_c(\alpha_l, P_l) \geq R_{\min,\, c}$ & $R_d(\alpha_l, P_l) < R_{\min,\, d}$, **then** $P_l^* = \frac{r_2}{m_3(1-\alpha_l) - m_4 r_2 \alpha_l}$
if $R_c(\alpha_l, P_l) < R_{\min,\, c}$ & $R_d(\alpha_l, P_l) < R_{\min,\, d}$, **then**
$P_l^* = \max\left\{\frac{r_1}{m_1\alpha_l - m_2 r_1(1-\alpha_l)}, \frac{r_2}{m_3(1-\alpha_l) - m_4 r_2 \alpha_l}\right\}$
if $R_c(\alpha_l, P_l) \geq R_{\min,\, c}$ && $R_d(\alpha_l, P_l) \geq R_{\min,\, d}$, **then** $P_l^* = P_l$

(2.7) Calculate $\eta_{EE}(\alpha_l, P_l^*)$ by (11)

3: $\eta_{EE}(P) = \max\left\{\left[\eta_{EE}(\alpha_1, P_1^*), \cdots, \eta_{EE}\left(\alpha_l, P_l^*\right), \cdots \eta_{EE}\left(\alpha_2, P_2^*\right)\right]\right\}$

4: $\eta_{EE}^P = \max\{[\eta_{EE}(P_{\min}), \cdots, \eta_{EE}(P)]\}$

---

4 Simulation Results

In this section, we evaluate the validity of the proposed scheme by the computer simulation. We consider an uplink DAS with N_t remote antennas. The polar coordinate of RA_1 is (0, 0), the $(N_t - 1)$ RA's polar coordinates are $\left(\sqrt{3/7}r,\ \pi i/3\right),\ i = 1, \cdots, N_t - 1$, where $N_t = 7$, $r = 1000$m is the radius of the cell. The location of CU is $(7r/10,\ \pi/3)$, the D2D-T is $(8r/10,\ \pi/2)$, and D2D-R is $(9r/10,\ \pi/2)$. The maximum total transmit power of CU and D2D-T is $P_{\max}$=2W. The noise power is $\sigma_n^2 = -70$dBm, the path loss factor is $\beta = 3$, the step is $\varepsilon_P = 0.05$ and the circuit power is $P_c = 5$W. For simplicity, we consider $R_{\min,\,c} = R_{\min,\,d} = R_{\min}$. The simulation results are illustrated in Figs. 2, 3 and 4, respectively.

In Fig. 2, the results of EE versus total power constraint $P_{\max}$ under different minimum rate constraints $R_{\min}$ are shown, where $\varepsilon_\alpha = 0.05$. We consider two minimum rate constraints $R_{\min}$=4 bit/s/Hz and $R_{\min}$=5 bit/s/Hz. Based on two different minimum rate constraints, we find that the EE with $R_{\min}$=4 bit/s/Hz is higher than that with $R_{\min}$=5 bit/s/Hz. The reason is that the increase of minimum rate constraint results in the increase of transmit power because of the requirement of higher rate, which brings about the decrease of EE. Besides, the EE increases firstly and then remains constant with the increase of $P_{\max}$. This is because when the maximum total power $P_{\max}$ is small, the CU's and DU's powers are limited due to the maximum power constraint. As $P_{\max}$ increases, however, both of them can achieve optimal power and the EE begins to increase as well. Besides, due to

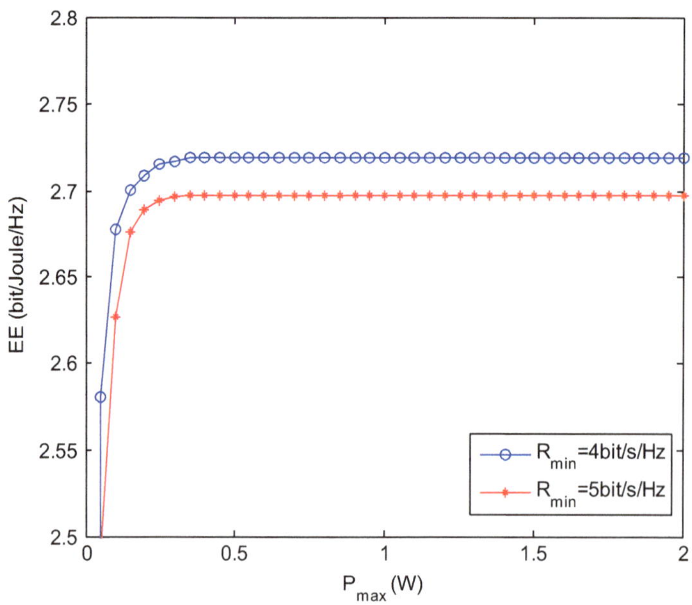

Fig. 2 The EE with different minimum rate constraints

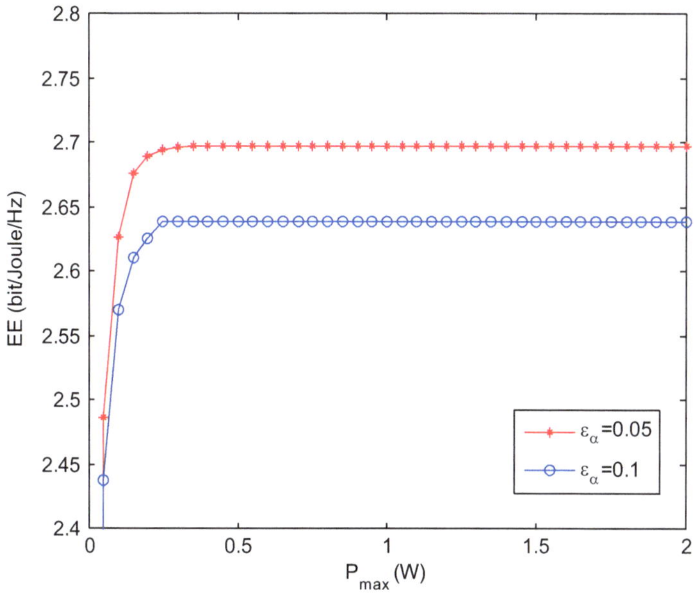

Fig. 3 The EE with different steps

the minimum rate constraints of CU and DU, the EE remains constant when $P_{\max}$ increases to a certain value.

Figure 3 shows the EE versus total power under different steps ε_α. In this simulation, we set the minimum rate constraint is $R_{\min} = 5$bit/s/Hz. As shown in Fig. 3, the EE with step $\varepsilon_\alpha = 0.05$ is higher than that with $\varepsilon_\alpha = 0.1$, as expected. The reason is that $\varepsilon_\alpha = 0.05$ has a higher degree of accuracy than $\varepsilon_\alpha = 0.1$, and more accurate value of power allocation can be attained. Therefore, the former can achieve higher EE than the latter.

From Fig. 4, it is found that the EE versus total power under different D2D locations, where $R_{\min} = 5$bit/s/Hz and $\varepsilon_\alpha = 0.05$. The polar coordinates of location 1 are $(8r/10,\ 5\pi/12)$ for D2D-T and $(9r/10,\ 5\pi/12)$ for D2D-R. The polar coordinates of location 2 are $(8r/10,\ \pi/2)$ for D2D-T and $(9r/10,\ \pi/2)$ for D2D-R. The polar coordinates of location 3 are $(8r/10,\ 7\pi/12)$ for D2D-T and $(9r/10,\ 7\pi/12)$ for D2D-R. Three different locations change the distance from DU to CU and the EE increases as the distance increases, as expected. This is because when DU is closer to CU, the interference between DU and CU increases and the transmit power decreases accordingly. Namely, CU and DU sharing the spectrum resources should stay away from each other when we design a distributed antenna system with D2D communication.

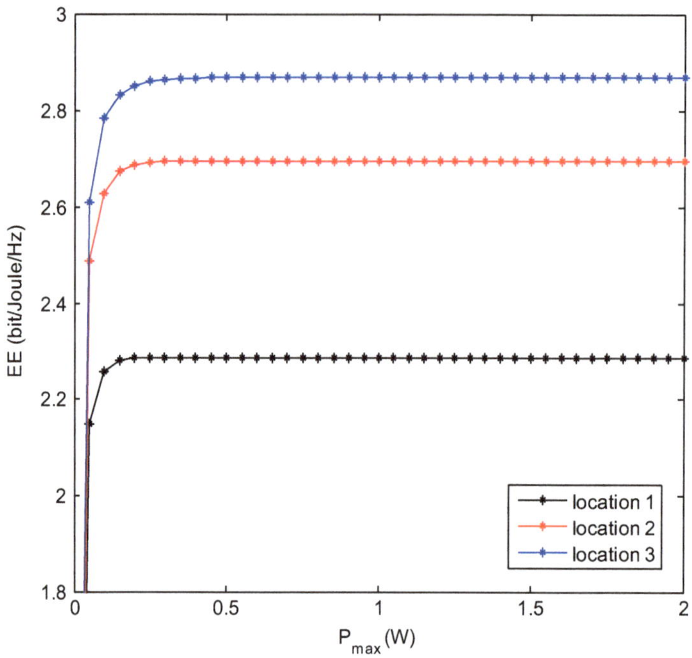

Fig. 4 The EE with different D2D locations

5 Conclusion

We have developed an energy-efficient power allocation scheme for D2D communication underlaying distributed antenna system. Compared with the existing research work, we present a D2D pair reuses the uplink channel resources of the DAS. To fulfill the maximum EE, we formulate the optimization problem subject to the maximum total power constraint and the minimum rate constraints of CU and DU. Due to the pseudo-concave of objective function in optimization problem, we propose an effective algorithm based on the gradient descent method and Armijo method to maximize the EE. Numerical results have demonstrated the validity of the proposed scheme and algorithm, and it can achieve superior EE performance.

Acknowledgment This work is supported by National Key Scientific Instrument and Equipment Development Project (Grant No. 2013YQ200607), Natural Science Foundation of Jiangsu Province in China (BK20181289), and Postgraduate Research and Practice Innovation Program of Jiangsu Province (KYCX19_0191).

References

1. Gandotra, P., Jha, R.K.: Device-to-device communication in cellular networks: a survey. J. Netw. Comput. Appl. **71**(C), 99–117 (2016)
2. Shafi, M., Molisch, A.F., Smith, P.J., et al.: 5G: a tutorial overview of standards, trials, challenges, deployment and practice. IEEE J. Sel. Areas Commun. **35**(6), 1201–1221 (2017)
3. Kim, H., Lee, S.R., Song, C.I., et al.: Optimal power allocation scheme for energy efficiency maximization in distributed antenna systems. IEEE Trans. Commun. **63**(2), 431–440 (2015)
4. Choi, W., Andrews, J.: Downlink performance and capacity of distributed antenna systems in a multicell environment. IEEE Trans. Wirel. Commun. **6**(1), 69–73 (2007)
5. Nader Tehrani, M., Uysal, M., et al.: Device-to-device communication in 5G cellular networks: challenges, solutions, and future directions. IEEE Commun. Mag. **52**(5), 86–92 (2014)
6. Yu, C., Doppler, K., Ribeiro, C., et al.: Resource sharing optimization for device-to-device communication underlaying cellular networks. IEEE Trans. Wirel. Commun. **10**(8), 2752–2763 (2011)
7. He, A., Wang, L., Chen, Y.: Spectral and energy efficiency of uplink D2D underlaid massive MIMO cellular networks. IEEE Trans. Commun. **65**(9), 3780–3793 (2017)
8. Wu, Y., Wang, J., Qian, L., et al.: Optimal power control for energy efficient D2D communication and its distributed implementation. IEEE Commun. Lett. **19**(5), 815–818 (2015)
9. Li, X., He, C., Huang, L., et al.: Energy efficient power allocation for co-located antenna systems with D2D communication. AEU – Int. J. Electron. Commun. **83**, 100–105 (2017)
10. Doppler, K., Rinne, M., Wijting, C., et al.: Device-to-device communication as an underlay to LTE-advanced networks. IEEE Commun. Mag. **47**(12), 42–49 (2009)
11. Zhu, D., Wang, J., Swindlehurst, A.L., et al.: Downlink resource reuse for device-to-device communications underlaying cellular networks. IEEE Signal Process. Lett. **21**(5), 531–534 (2014)

A Cluster-Based Interference Management with Successive Cancellation for UDNs

Lihua Yang, Junhui Zhao, Feifei Gao, and Yi Gong

1 Introduction

In order to satisfy the requirements of the wireless communication, the fifth generation (5G) communication networks have been widely developed in the following three aspects: spectral efficiency, spectrum expansion and network densification [1]. As one of the promising approaches to improve the system spectral efficiency, massive multiple-input multiple-output (MIMO) has the capability to enhance the 5G network reliability [2]. Providing Gigahertz transmission bandwidth, millimeter-wave (mmWave) can effectively expand the spectrum resources [3]. As the core characteristics of 5G cellular networks, the ultra-dense network (UDN) has appeared to meet the explosive capacity of mobile communication systems [4]. Consists of plenty of macro base stations (MBSs) and many types of small base stations (SBSs), the UDN is developed from the heterogeneous network (HetNet) to meet the demands of the high mobile data volume [5]. However, the increasing number and the decreasing size of base stations (BSs) result in the complicated topology of

L. Yang
School of Electronic and Information Engineering, Beijing Jiaotong University, Beijing, China

J. Zhao (✉)
School of Electronic and Information Engineering, Beijing Jiaotong University, Beijing, China

School of Information Engineering, East China Jiaotong University, Nanchang, China

F. Gao
Tsinghua National Laboratory for Information Science and Technology (TNList), Beijing, China

Y. Gong
Department of Electrical and Electronic Engineering, Southern University of Science and Technology, Shenzhen, China

B. Ye et al. (eds.), *2nd International Conference on 5G for Ubiquitous Connectivity*, EAI/Springer Innovations in Communication and Computing,
https://doi.org/10.1007/978-3-030-22316-8_5

networks and serious interference. Therefore, the interference management plays a more and more important role in improving the performance of network.

Previous researchers have provided an overview of interference management scheme in cellular networks [6–9], e.g., cell range expansion, enhanced inter-cell interference coordination, cognitive interference management, and so on. However, these studies only focussed on decreasing the cross-tier interference and there was little advantage of co-tier interference mitigation in the network. Unfortunately, some of them possessed high computational complexity which made it difficult to be applied in UDNs. Recently, many researchers have paid more attention to reducing co-tier interference and computational complexity. Clustering is considered to be one of the promising ways to reduce the co-tier interference and make the topology structures of UDN be simplified.

Cluster-based interference management schemes have been studied in many literatures [10–14]. In the literature [10], authors designed a distributed interference management scheme in two-tier networks. Abdelnasser et al. proposed a clustering algorithm to decrease the co-tier interference in FBSs [11]. However, the authors ignored the quality of service (QoS) of users, and these two schemes were unable to guarantee the high transfer rate of users in the network. Wei et al. [12] studied a cluster-based wireless resource management which was split into a modified K-means algorithm and a supplementary allocation algorithm in UDN. The work in [13] presented a cluster-based energy-efficient resource allocation scheme to improve the energy-efficient of the network. To improve the throughout of FBS networks, Dai et al. [14] proposed an interference management scheme based on joint clustering and resource allocation with an acceptable complexity. Nevertheless, the above interference management schemes did not consider the interference among users (i.e., intra-cluster interference) which was serious in hot spot region. Nam et al. [15] indicated that the interference among users can be decreased by advanced receivers (i.e., successive interference cancellation) with interference joint detection or decoding. Hence, to decrease all types of interference simultaneously in UDN, we are looking forward to designing an effective interference management scheme.

The main contribution of this paper can be summarized as follows: 1) Considering all types of interference, we formulate an optimization problem based on joint clustering, subchannel allocation and successive cancellation among users. 2) To solve this optimization problem, we propose a clustering algorithm based on interference graph which can distribute FBSs and femto user equipments (FUEs) into disjoint clusters and groups synchronously. 3) Then, a subchannel allocation algorithm, which aims to allot the subchannel to each FBS cluster, is provided to minimize the cross-tier interference. 4) Moreover, detecting and demodulating successively the largest received power of FUE, a successive interference cancellation (SIC) detection algorithm is developed to reduce the interference among users in the same FUE group. Numerical results validate the effectiveness and efficiency of our proposal.

2 Model and Formulation

2.1 System Model

A two-tier downlink UDN with densely deployed FBSs operating within a MBS is considered in this paper. All MBSs and FBSs are located with the same radius R_m and R_f, respectively. Users are distributed randomly within the coverage of the MBS. To allow all users to access flexibly, all FBSs are assumed to adopt the open subscriber group (OSG) configuration. Furthermore, it is assumed that the frequency reuse factor is equal to one. Figure 1 shows the topology under consideration in this paper, and it demonstrates that there are two types of interference (cross-tier interference and co-tier interference) in the downlink. Our hypothesis also contains that all users in the network are capable of identifying the interference source [16].

A user is considered to be a macro user equipment (MUE) only if the signal power received from MBSs is greater than FBSs, and vice versa, which is called the maximal received power association. We utilize signal to interference-plus-noise ratio (SINR) to evaluate the performance of the network. Moreover, we assume that all MBSs and FBSs transmit in the same frequency (i.e., co-channel deployment). The SINR of MUE m served by MBS l using subchannel n is given by

$$\mathrm{SINR}_{l,m}^{(n)} = \frac{P_l \beta_{l,m}^{(n)} \varphi_{l,m}}{I_M + I_F + I_U + N_0 B}, \tag{1}$$

where P_l represents the transmission power of BS, $\beta_{l,m}^{(n)}$ is the fast fading power from BS l to user m in subchannel n, $\varphi_{l,m}$ is the path loss attenuation factor from

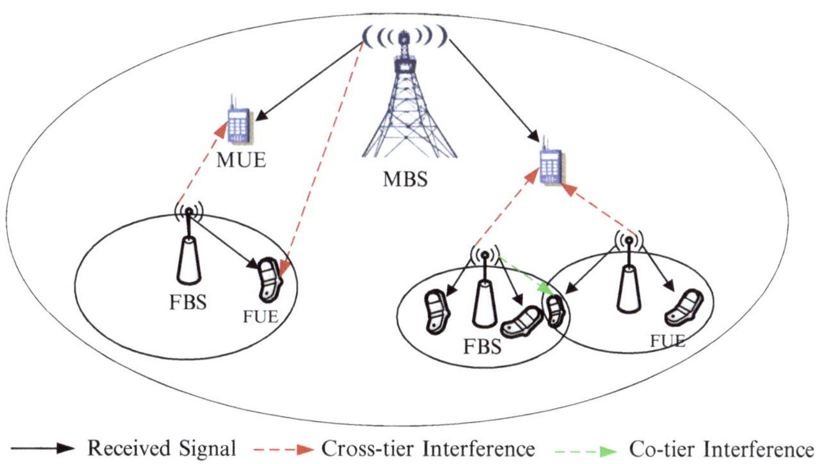

Fig. 1 Network topology under consideration

BS l to user m, $I_M = \sum_{\tau \neq l, \tau \in \Phi} P_\tau \beta_{\tau,m}^{(n)} \varphi_{\tau,m}$ describes the interference from other MBSs, $I_F = \sum_{\kappa \in \Theta_l} P_{\kappa(l)} \beta_{\kappa(l),m}^{(n)} \varphi_{\kappa(l),m}$ is the total interference from all FBSs in MBS l, $\Phi = \{1, 2, \cdots, L\}$ and $\Theta = \{\Theta_1, \Theta_2, \cdots, \Theta_L\}$ are the set of all MBSs and FBSs in the network respectively, Θ_l denotes the set of FBSs located in MBS l, $I_U = \sum_{m' \in \mathbb{U}_l, m' \neq m} P_{m'} \beta_{l,m'}^{(n)} \varphi_{m,m'}$ is the interference from other MUEs served by the same MBS with user m, $\mathbb{U} = \{\mathbb{U}_1, \mathbb{U}_2, \cdots, \mathbb{U}_L\}$ denotes the set of all MUEs in the network, $\mathbb{U}_l$ is the set of MUEs in MBS l, $N_0 = -174dBm/Hz$ is the noise power spectral density, and B is the transmission bandwidth.

Similarly, the corresponding SINR of FUE j associated with FBS f in the coverage of MBS l is expressed as

$$\mathrm{SINR}_{f(l),j}^{(n)} = \frac{P_{f(l)} \beta_{f(l),j}^{(n)} \varphi_{f(l),j}}{I_M + I_F + I_U + N_0 B}, \tag{2}$$

where $I_M = \sum_{\tau \in \Phi} P_\tau \beta_{\tau,j}^{(n)} \varphi_{\tau,j}$ is the cross-tier interference from all MBSs, $I_F = \sum_{\kappa \neq f, \kappa \in \Theta_l} P_{\kappa(l)} \beta_{\kappa(l),j}^{(n)} \varphi_{\kappa(l),j}$ is the co-tier interference from all FBSs except the serving FBS in MBS l, and $I_U = \sum_{j' \in \mathbb{F}_{f(l)}, j' \neq j} P_{j'} \beta_{l,j'}^{(n)} \varphi_{j,j'}$ is the interference from other FUEs associated with the same FBS in MBS l, $\mathbb{F} = \{\mathbb{F}_1, \mathbb{F}_2, \cdots, \mathbb{F}_L\}$ is the set of all FUEs in the network, $\mathbb{F}_l = \left\{\mathbb{F}_{1(l)}, \mathbb{F}_{2(l)}, \cdots, \mathbb{F}_{D(l)}\right\}$ is the set of FUEs in MBS l.

Accordingly, we can obtain the capacity of MUE m and FUE j in subchannel n and the expressions are

$$\mathrm{R}_{l,m}^{(n)} = \xi_{l,m} \times \log_2(1 + \mathrm{SINR}_{l,m}^{(n)}) \tag{3}$$

and

$$\mathrm{R}_{f(l),j}^{(n)} = \eta_{l,j} \times \log_2(1 + \mathrm{SINR}_{f(l),j}^{(n)}), \tag{4}$$

where $\xi_{l,m} = {}^{B}/_{U_l}$ and $\eta_{l,j} = {}^{B}/_{F_l}$ are the bandwidths of the MUE and FUE, respectively, U_l is the number of MUEs in MBS l, and F_l is the number of FUEs in MBS l.

2.2 Problem Formulation

Jointing clustering, subchannel allocation and successive cancellation among users, we formulate the following optimization problem to maximize the system capacity,

$$\begin{aligned}
&\max \sum_{l=1}^{L}\sum_{m\in\mathbb{U}_l}\sum_{n=1}^{N}\rho_{l,m}^{(n)}\mathrm{R}_{l,m}^{(n)}+\sum_{l=1}^{L}\sum_{i=1}^{|\mathcal{C}_l|}\sum_{f\in\mathcal{C}_{i(l)}}\sum_{j\in\mathbb{F}_{f(l)}}\sum_{n=1}^{N}\rho_{f(l),j}^{(n)}\mathrm{R}_{f(l),j}^{(n)}\\
&s.t.\ \mathrm{C1}:\sum_{n=1}^{N}\rho_{l,m}^{(n)}R_{l,m}^{(n)}\geqslant R_{m,\min},\forall l,m,\\
&\quad \mathrm{C2}:\sum_{n=1}^{N}\rho_{f(l),j}^{(n)}\mathrm{R}_{f(l),j}^{(n)}\geqslant R_{j,\min},\forall l,f,j,\\
&\quad \mathrm{C3}:\sum_{m\in\mathbb{U}_l}\rho_{l,m}^{(n)}=1,\ \sum_{j\in\mathbb{F}_{f(l)}}\rho_{f(l),j}^{(n)}=1,\forall l,f,n,\\
&\quad \mathrm{C4}:\rho_{l,m}^{(n)}\in\{0,1\},\forall l,m,n,\\
&\quad \mathrm{C5}:\rho_{f(l),j}^{(n)}\in\{0,1\},\forall l,f,j,n,\\
&\quad \mathrm{C6}:\bigcup_{l=1}^{L}\bigcup_{i=1}^{|\mathcal{C}_l|}\mathcal{C}_{i(l)}=\mathcal{C},\ \bigcap_{i=1}^{|\mathcal{C}_l|}\mathcal{C}_{i(l)}=\emptyset,
\end{aligned}\tag{5}$$

where $\rho_{l,m}^{(n)}$ and $\rho_{f(l),j}^{(n)}$ can only be either 0 or 1 indicating whether the nth subchannel is occupied by a user or not, $R_{m,\min}$ and $R_{j,\min}$ are the minimal rate requirements of user m and j respectively, $\mathcal{C}=\{\mathcal{C}_1,\mathcal{C}_2,\cdots,\mathcal{C}_L\}$ is the set of total FBS clusters in the network, $\mathcal{C}_l=\left\{\mathcal{C}_{1(l)},\mathcal{C}_{2(l)},\cdots,\mathcal{C}_{i(l)},\cdots\right\}$ is the set of FBS clusters in MBS l, and $\mathcal{C}_{i(l)}$ is the ith FBS cluster set in MBS l.

In the optimization problem (5), the objective is to maximize the system capacity of the network subject to data rate requirement $R_{m,\min}$ and $R_{j,\min}$ as indicated in C1 and C2. C3, C4 and C5 are the exclusion constraints indicating that subchannel n can be only used in one MBS or FBS, respectively. Constraints C6 indicates that the set of FBS clusters $\mathcal{C}_l$ of all MBSs form the entire cluster $\mathcal{C}$ in the network and the arbitrary two FBS clusters in a MBS are disjoint.

3 Interference Management Scheme

Involving variables $\mathcal{C}_{i(l)}$, $\mathcal{C}_l$ and binary variables $\rho_{l,m}^{(n)}$, $\rho_{f(l),j}^{(n)}$, problem (5) is complicated to obtain the optimal solution. To obtain the solution, we divide it into three procedures: clustering algorithm, subchannel allocation and SIC detection algorithm. In the clustering algorithm, all FBSs and FUEs in the network are divided into several clusters and groups simultaneously based on interference graph to avoid the co-tier interference. Meanwhile, the cross-tier interference can be reduced via the subchannel allocation algorithm which distributes different subchannels to all FBS clusters. Moreover, we use SIC detection algorithm to reduce the interference among users in the same group. The detailed procedures of the proposed algorithm will be shown in the following sub-sections.

3.1 Structure of Interference Graph

Given interference graph $G(E, W)$ with nodes (from set E) and edge weight $w(a, b)$ for each edge (a, b) (from set W), we regard users as nodes in the graph. Denoting the edge weight as the interference degree, the interference between users is simplified to that between nodes. Moreover, W is the bi-directional edge set and each element possess a non-negative weight. Hence, the equation $w(a, b) = w(b, a)$ can be established. We assume that there are 5 users, and the interference graph is shown in Fig. 2. The critical factor to structure the interference graph is to calculate the edge weights. The detailed processes are illustrated in the following.

1. First, we obtain SINRs of FUEs which constitute the interference graph.
2. Then, taking user u as the target and v as the interference, the edge weight between two users is calculated by

$$\lambda_{u,v} = \frac{1 + {}^{1}/\text{SINR}_u}{1 + {}^{1}/\text{SINR}_v}, \tag{6}$$

3. Similar to (6), we can obtain the expression of $\lambda_{v,u}$. The edge weight between user u and v is represented as

$$w\left(u, v\right) = w\left(v, u\right) = \max\left(\lambda_{u,v}, \lambda_{v,u}\right). \tag{7}$$

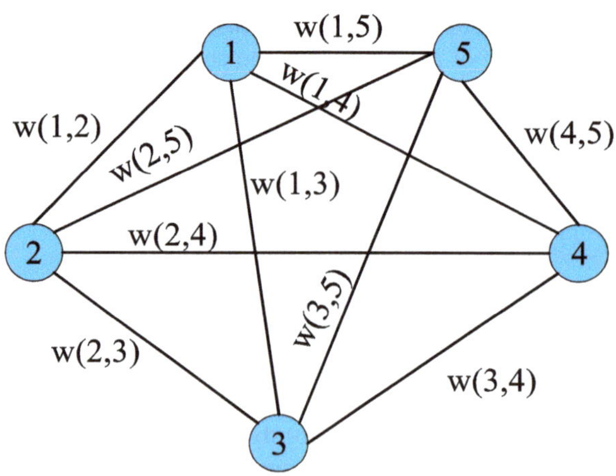

Fig. 2 The interference graph constructed of users

3.2 Clustering Algorithm

After obtain the edge weights, we compute some necessary variables first.

1. Calculate the sum of edge weights of user u as

$$W_u = \sum\nolimits_{v=1, v \neq u}^{F_l} w(u, v), \tag{8}$$

2. Define the harmonic mean of distance d_u as

$$d_u = \frac{1}{\sum_{u=1, u \neq v}^{F_l} {d_{u,v}}^{-1}}, \tag{9}$$

where $d_{u,v}$ is the distance between user u and v.

3. Combining W_u and d_u, the combination weight is shown as

$$W_{update} = k1 \times W_u + k2 \times A \times d_u, \tag{10}$$

where k_1 and k_2 are weight factors, and the sum of two values equals to 1.

We assume that $\mathbb{GU} = \{\mathbb{GU}_1, \mathbb{GU}_2, \ldots, \mathbb{GU}_L\}$ denotes the set of FUE groups in the network and each element in $\mathbb{GU}$ represents a set of FUE groups in a MBS. Our clustering algorithm is described detailedly in Algorithm 1.

3.3 Subchannel Allocation

After accomplishing the clustering algorithm, we have obtained the FBS clusters and FUE groups to reduce the constraint conditions in (5) and the optimization problem can be rewritten as

$$\max \sum_{l=1}^{L} \sum_{m \in \mathbb{U}_l} \sum_{n=1}^{N} \rho_{l,m}^{(n)} \mathrm{R}_{l,m}^{(n)} + \sum_{l=1}^{L} \sum_{i=1}^{|C_l|} \sum_{f \in c_{i(l)}} \sum_{j \in \mathbb{F}_{f(l)}} \sum_{n=1}^{N} \rho_{f(l),j}^{(n)} \mathrm{R}_{f(l),j}^{(n)} \tag{11}$$

$$s.t.\ \mathrm{C1}, \mathrm{C2}, \mathrm{C3}, \mathrm{C4}, \mathrm{C5}, \forall l, m, f, j, n$$

Our goal is to find the channel assignment set $\mathbb{Y} = \{\mathbb{Y}_1, \mathbb{Y}_2, \ldots, \mathbb{Y}_L\}$ for clusters in the network, where $\mathbb{Y}_l = \left[y_{i,n}\right]$ and $y_{i,n}$ is equal to one if subchannel n is assigned to ith cluster $C_{i(l)}$ in MBS l and zero, otherwise. Let $\Omega = \{\Omega_1, \Omega_2, \ldots, \Omega_L\}$ be the subchannel pool, where $\Omega_l = \{1, 2, \ldots N\}$. To obtain the optimal subchannel for each cluster, the subchannel allocation algorithm used in this paper is exhaustive search instead of the algorithm in [16], where calculate the sum rate of each cluster, assign the optimal n^* in Ω_l to the corresponding cluster, and repeat several times until all clusters are allocated with subchannels.

Algorithm 1 Clustering algorithm

Input: W_{update}, w_2, c, $w(u, v)$, $\mathbb{F}$, $\mathbb{V}$(intermediate variable);
Output: $\mathbb{GU}$, $\mathcal{C}$;
Initialize: U_{num}, w_{th}, $\mathbb{V} = \mathbb{F}$, $|\mathbb{GU}| = c$, w_2 and c are contents, $\mathbb{GU}$ and $\mathcal{C}$ are sets of FUE groups and FBS clusters;
for $l = 1 : L$ **do**
  Calculate w_{th} according to $w_{th} = \left|\mathcal{C}_l\right| \times w_2$;
  Sort W_{update} of in ascending order to get the array U_{num};
  for $i = 1 : (|\mathbb{GU}_l| - 1)$ **do**
    Select the smallest element $U_{\min}$ in array U_{num} and remove it from U_{num} ;
    Regard the corresponding user u as the ith FUE group head and let $\mathbb{GU}_{i(l)} = \mathbb{GU}_{i(l)} \cup \{u\}$, $\mathbb{F}_l = \mathbb{F}_l \backslash \{u\}$;
    for $v = 1 : |\mathbb{V}_l|$ **do**
      $\mathbb{V}_l \leftarrow \mathbb{V}_l \backslash \{u\}$;
      if $w\left(u, \mathbb{V}_l\left(v\right)\right) \geqslant w_{th}$ **then**
        Add $\mathbb{V}_l\left(v\right)$ into $\mathbb{GU}_{i(l)}$ as $\mathbb{GU}_{i(l)} \leftarrow \mathbb{GU}_{i(l)} \cup \{\mathbb{V}_l\left(v\right)\}$, remove $\{\mathbb{V}_l\left(v\right)\}$ from $\mathbb{V}_l$ as $\mathbb{V}_l \leftarrow \mathbb{V}_l \backslash \{\mathbb{V}_l\left(v\right)\}$, and remove the corresponding W_{update} from U_{num};
      else
        $v = v + 1$;
      end
    end
  end
  Put the surplus elements of set $\mathbb{V}_l$ into the last group, and then form the previous FUE group $\mathbb{GU}_l$.
end
for $l = 1 : L$ **do**
  for $i = 1 : \left|\mathbb{GU}_l\right|$ **do**
    Search the served FBS for each FUE in the group $\mathbb{GU}_{i(l)}$ successively to form the corresponding set of FBS cluster $\mathcal{C}_{i(l)}$;
    Search FUEs served by FBSs in cluster $\mathcal{C}_{i(l)}$ and put them into the corresponding group to update the FUE group $\mathbb{GU}_{i(l)}$;
  end
  Delete the repetitive FBSs and FUEs of all clusters and groups in MBS l to form the final FBS cluster $\mathcal{C}_l$ and FUE group $\mathbb{GU}_l$.
end

3.4 SIC Detection Algorithm

When users utilize the non-orthogonal multiple access, the information received by the downlink user contains multiple access interference from other users. The idea of SIC detection is to strip the user data successively. We assume the set $\mathbb{P} = \{\mathbb{P}_1, \mathbb{P}_2, \ldots, \mathbb{P}_L\}$ stores the power values of all FUEs, and $\mathbb{I} = \{\mathbb{I}_1, \mathbb{I}_2, \ldots, \mathbb{I}_L\}$ represents the interference from other FUEs in the same group. The details are given in Algorithm 2.

Algorithm 2 SIC detection algorithm

Input: $\mathbb{GU}$, $\mathbb{P}$, $\mathbb{F}$;
Output: $\mathbb{I}$;
Initialize: $\zeta = \emptyset$, $\mathbb{I} = \emptyset$;
for $l = 1 : L$ **do**
  for $i = 1 : |\mathbb{GU}_l|$ **do**
    Select the maximal power of users in $\mathbb{G}_{i(l)}$;
    Regard the corresponding user as u^*, then delete the power of u^* and add user u^* into ζ as $\zeta \leftarrow \zeta \cup \{u^*\}$;
    while $\{u\} \in \mathbb{GU}_{i(l)}$ **do**
      Reselect the maximal power, and add the corresponding user v into ζ as $\zeta \leftarrow \zeta \cup \{v\}$;
      Calculate the interference among users by $I_U = \sum_{m' \in \mathbb{GU}_{i(l)} \setminus \zeta} P_{m'} \beta^{(n)}_{l,m'} \varphi_{m,m'}$ and put it into $\mathbb{I}_l$ as $\mathbb{I}_l \leftarrow \mathbb{I}_l \cup \{I_U\}$;
      Remove $P_{\max}$ from $\mathbb{P}_{i(l)}$.
    end
  end
end

We analyze the complexity of our proposed scheme in different stages. Assuming the average number of clusters and FBSs located in each cluster in a MBS are $\mathcal{C}_g$ and B_g, respectively. Moreover, the assumptions also contain that the average number of FUE groups in a MBS is $\mathbb{GU}_g$, and the average number of FUEs located in each group is F_g. In the worst case, the clustering algorithm has a complexity of $\mathcal{O}\left(F_l^2 + F_l + \mathcal{C}_g B_g + \mathbb{GU}_g F_g\right)$. The computational complexity of subchannel allocation algorithm is $\mathcal{O}\left(\mathcal{C}_g!\right)$ in the worst case. The computational complexity of SIC detection algorithm is $\mathcal{O}\left(\mathbb{GU}_g F_g^2\right)$. With the accepted complexity, the proposed interference management scheme effectively reduces the co-tier interference by clustering algorithm, cross-tier interference by subchannel allocation algorithm and the interference among users in the same group by SIC detection algorithm.

4 Simulation Results

In this section, we conduct the simulations to verify the effectiveness of the proposed interference management scheme in UDN. Considering a scenario with plenty of MBSs and FBSs, we utilize Reyleigh fading to model the channels between BSs and users. The path loss between MBS and MUE is 128.1+37.6*$\log_{10} D$, and it is 140.7+36.7*$\log_{10} d$ between FBS and FUE. Other simulation parameters are shown in Table 1. We evaluate the performance of our proposed scheme in comparison with the following two schemes: the optimal FBS subchannel allocation (OFBSSA) and cluster-based FBS subchannel allocation (CFBSSA).

Table 1 Simulation parameters

Parameter	Value	Parameter	Value
Number of MBS L	19	Total bandwidth B	10 MHz
Transmit power of MBS	46 dBm	Transmit power of FBS	23 dBm
Coverage area of MBS R_m	500 m	Coverage area of FBS R_f	20 m

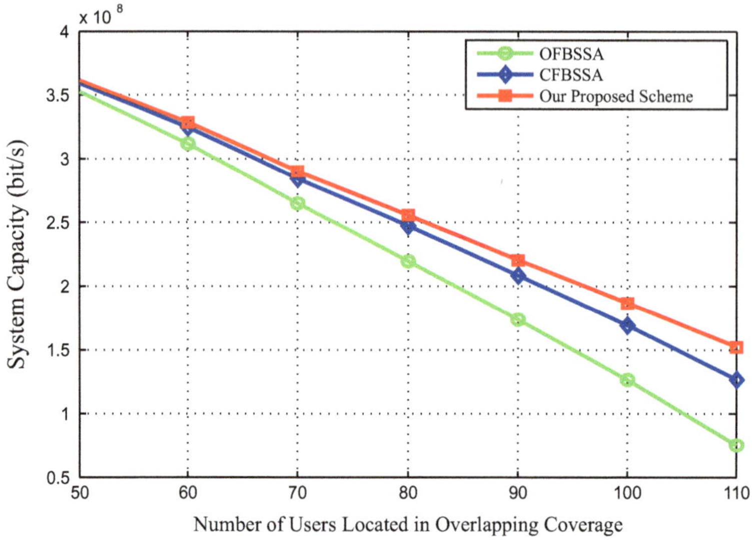

Fig. 3 System capacity vs users

Figure 3 shows the simulation results between the system capacity and the number of users located in overlapping region of two FBSs. The holistic tendency of the figure shows that the system capacity decreases with the increasing number of users. The main reason is that the interference increases with the growth of the edge users in FBSs. Hence, the system capacity decreases sharply. Comparing the three curves in the graph, it can be found that the capacity of the proposed scheme is better than the other two options when the number of users in the overlapping region is communal. The advantage is more significant when the number of edge users is big enough. Moreover, Fig. 3 shows that the proposed scheme reduces the interference of edge users and improves the system capacity.

Figure 4 is about the performance of three schemes in Fig. 3. We can see that the proposed scheme has great preponderance in improving the system capacity, especially when there are plenty of edge users. When the number of users is fixed, the percentage of the OFBSSA and our proposed scheme is higher than the CFBSSA and our proposed scheme. With a growing number of users located in overlapping region, it is clear that the improved percentage becomes higher. Furthermore, the advantage of the proposed scheme is gradually distinct and it is in a position to meet the capacity requirements of users in hotspot areas.

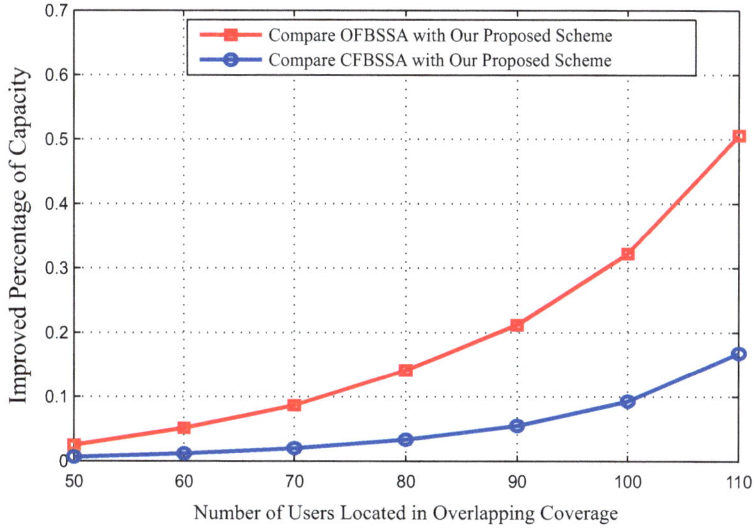

Fig. 4 Improved percentage vs users

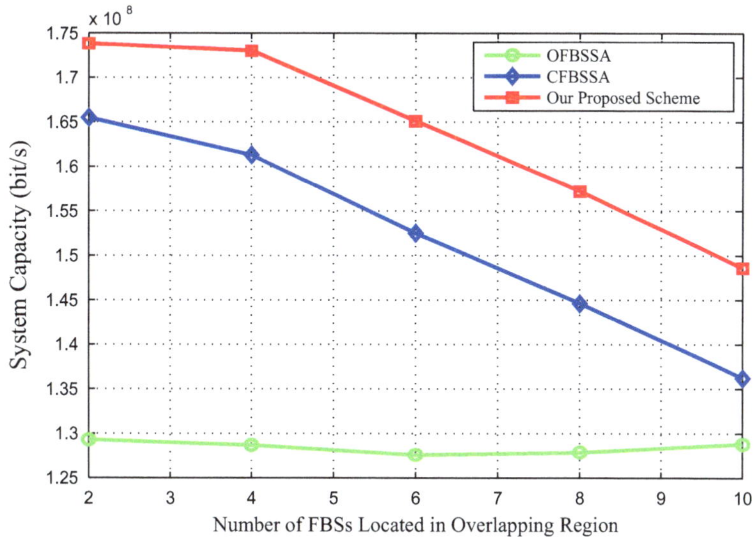

Fig. 5 System capacity vs FBSs

Figure 5 describes the relationship between the number of FBSs located in overlapping region and the system capacity. The result shows that the system capacity is reduced with the increasing number of FBSs owning the common coverage. When the total number of FBSs is fixed, the maximal received power of users are constants. However, the interference becomes more and more serious

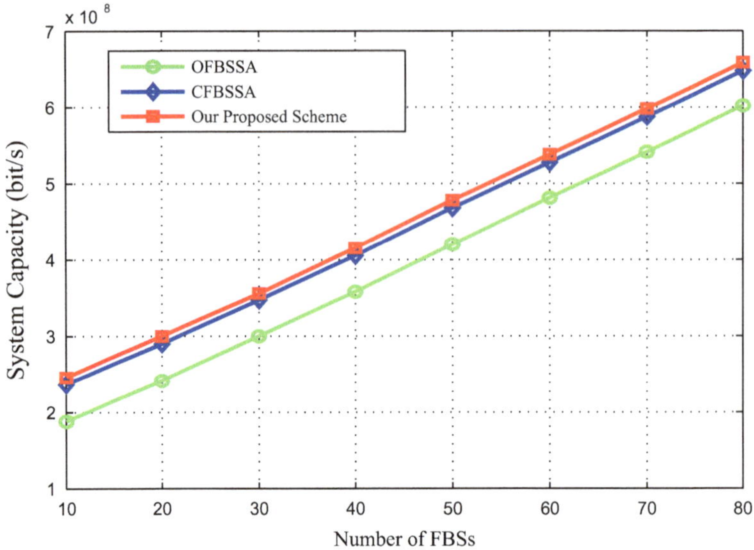

Fig. 6 System capacity vs the density of FBSs

as the number of FBSs located in overlapping region increases. Therefore, the three curves show a decreasing trend in Fig. 5. Nevertheless, the system capacity of the proposed scheme is larger than the other two schemes clearly. Hence, the proposed scheme shows a certain application value in specific dense areas.

We also investigate the system capacity with the density of FBSs in Fig. 6. On the one hand, it indicates that the system capacity increases as the density of FBSs increases, which is caused by the face that FUEs have more choices to associate with the FBS providing a higher power. Moreover, the co-tier interference is reduced with the increasing number of FBSs compared with OFBSSA and CFBSSA. The SIC detection algorithm can decrease the interference among FUEs in the same group contrasting of CFBSSA scheme and our proposed scheme. On the other hand, compared with the other two schemes, the capacity of the proposed scheme is significantly improved when the number of FBSs is a constant.

Finally, we compare the spectral efficiency of the three schemes by varying the number of FBSs and the simulation result is shown in Fig. 7. As the number of FBS increases, the spectral efficiency of the system increases for three schemes in Fig. 7. The reason is that the proposed scheme can obviously improve the SINR of all users, which results in the increasing of spectral efficiency under the average bandwidth allocation condition. Moreover, it is obvious that our proposed scheme has a greater promotion in improving the spectral efficiency, which is significative for the future network.

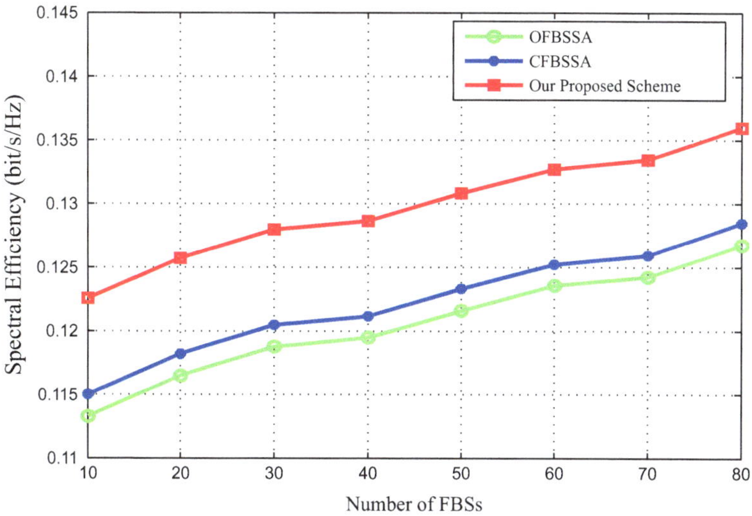

Fig. 7 Spectrum efficiency vs the density of FBSs

5 Conclusions

In this paper, we have proposed an interference management scheme in UDN which has been verified to be an effective scheme to improve the system capacity and spectral efficiency. The scheme has three stages, namely, clustering algorithm, subchannel allocation and SIC detection algorithm. In the clustering algorithm, to avoid the co-tier interference, we have divided the FBSs and FUEs into different clusters and groups respectively based on the interference graph. In the subchannel allocation stage, we have utilized a exhaustive search to reduce the cross-tier interference. Moreover, the interference among users in the same group has been decreased by using the SIC detection algorithm. Simulation results show that the system capacity and spectral efficiency have been improved compared with OFBSSA scheme and CFBSSA scheme. Furthermore, the proposed interference management scheme has a certain application prospect in 5G communication.

Acknowledgements This work is supported by the Fundamental Research Funds for the Central Universities (No. 2019YJS009), the National Natural Science Foundation of China (61661021), National Science and Technology Major Project of the Ministry of Science and Technology of China (2016ZX03001014-006), the Open Research Fund of National Mobile Communications Research Laboratory, Southeast University (No. 2017D14), Beijing Jiaotong University, Shenzhen Science and Technology Program under Grant (No. JCYJ20170817110410346).

References

1. Bhushan, N., Li, J., Malladi, D., et al: Network densification: the dominant theme for wireless evolution into 5G. IEEE Commun. Mag. **52**(2), 82–89 (2014)
2. Ni, S., Zhao, J., Gong, Y.: Optimal pilot design in massive MIMO systems based on channel estimation. IET Commun. **11**(7), 975–984 (2017)
3. Wang, B., Gao, F., Jin, S., et al: Spatial and frequency wideband effects in millimeter-wave massive MIMO systems. IEEE Trans. Signal Process. **66**(13), 3393–3406 (2018)
4. Niu, C., Li, Y., Hu, R.Q., et al: Fast and efficient radio resource allocation in dynamic ultra-dense heterogeneous networks. IEEE Access **5**, 1911–1924 (2017)
5. Ge, X., Tu, S., Mao, G., et al: 5G ultra-dense cellular networks. IEEE Wirel. Commun. **23**(1), 72–79 (2016)
6. Zhou, Y., Liu, L., Du, H., et al: An overview on intercell interference management in mobile cellular networks: from 2G to 5G. In: 2014 IEEE International Conference on Communication Systems (ICCS), Macau, China, pp. 217–221 (2014)
7. Ni, S.J., Zhao, J.H., Yang, H.H., Quek, T.Q.S., Gong, Y.: Small cell range expansion with interference mitigation for downlink massive MIMO HetNets. In: 2018 IEEE Global Communications Conference (GLOBECOM) (2018)
8. Lopez-Perez, D., Guvenc, I., Roche, G.D.L., et al.: Enhanced intercell interference coordination challenges in heterogeneous networks. IEEE Wirel. Commun. **18**(3), 22–30 (2011)
9. Kaimaletu, S., Krishnan, R., Kalyani, S., et al.: Cognitive interference management in heterogeneous femto-macro cell networks. In: 2011 IEEE International Conference on Communications (ICC), Kyoto, Japan, pp. 1–6 (2011)
10. Hosseini, K., Dahrouj, H., Adve, R.: Distributed clustering and interference management in two-tier networks. In: 2012 IEEE Global Communications Conference (GLOBECOM), Anaheim, CA, USA, pp. 4267–4272 (2012)
11. Abdelnasser, A., Hossain, E., Dong, I.K.: Clustering and resource allocation for dense FBSs in a two-tier cellular OFDMA network. IEEE Trans. Wirel. Commun. **13**(3), 1628–1641 (2014)
12. Wei, R., Wang, Y., Zhang, Y.: A two-stage cluster-based resource management scheme in ultra-dense networks. In: 2014 IEEE International Conference on Communications in China (ICCC), Shanghai, China, pp. 738–742 (2014)
13. Liang, L., Wang, W., Jia, Y., et al: A cluster-based energy-efficient resource management scheme for ultra-dense networks. IEEE Access **4**, 6823–6832 (2016)
14. Dai, J., Wang, S.: Clustering-based interference management in densely deployed femtocell networks. Digital Commun. Netw. **2**(4), 175–183 (2016)
15. Nam, W., Bai, D., Lee, J., Kang, I.: Advanced interference management for 5G cellular networks. IEEE Commun. Mag. **52**(5), 52–60 (2014)
16. Chang, Y.J., Tao, Z., Zhang, J., et al.: A graph-based approach to multi-cell OFDMA downlink resource allocation. In: 2008 IEEE Global Telecommunications Conference, New Orleans, LO, USA, pp. 1–6 (2008)

Delay Sensitive Application Partitioning and Task Scheduling in Mobile Edge Cloud Prototyping

Abdullah Lakhan, Dileep Kumar Sajnani, Muhammad Tahir, Muhammad Aamir, and Rakhshanda Lodhi

1 Introduction

Nowadays, proliferation of the mobile devices and computation demands of delay sensitive applications is becoming ubiquitous [1]. The MEC augment the capabilities of resource-constraint mobile devices and enable them to perform any compute intensive type application to execute. An offloading system is a technique in MEC which moves compute intensive tasks of mobile application to the edge or remote cloud for further processing. The foremost cost of the offloading system is computation and communication cost. The delay sensitive application can be modeled as a call graph. It is a challenge to partition the application based on those factors that degrade the QOS of the application, and efficient task scheduling such that average total time can be minimized.

In this paper, we are formulating the application partitioning and task scheduling problem for delay sensitive applications in mobile edge cloud environment. However, the application partitioning problem allows us to execute compute intensive application on the mobile device and offload heavy tasks from limited constraint mobile onto the cloud server for the execution. In this way, the quality of user experience is satisfied. On the other hand, after the application partitioning the divided tasks is executed such that application hard constraint could not be

A. Lakhan (✉) · D. K. Sajnani · R. Lodhi
School of Computer Science and Engineering, Southeast University, Nanjing, China
e-mail: abdullah@seu.edu.cn

M. Tahir
School of Software Technology, Dalian University of Technology, Dalian, China

M. Aamir
College of Computer Science, Sichuan University, Chengdu, China

B. Ye et al. (eds.), *2nd International Conference on 5G for Ubiquitous Connectivity*, EAI/Springer Innovations in Communication and Computing,
https://doi.org/10.1007/978-3-030-22316-8_6

compromised, to cope up with challenging task scheduling is efficient to execute all tasks on appropriate resources. Furthermore, the application can be represented as a call graph in which each node represented a task and each edge shows the data dependencies between tasks. The preliminary constraint for the application is that it must be executed within a given deadline.

MEC is the combination of three different technologies such as, mobile computing, wireless technology and cloud computing paradigm. However, In the literature, many architectures were proposed such as MAUI [1], CloneCloud [2], Cuckoo [3], ThinkAir [4] to achieved their objective, the primary objective was to offload the heavy computation of the application to the cloud server so that minimized mobile computation power and boosted the application performance. However, However, 80% compute intensive tasks of the mobile application have to be offloaded to the cloud server for execution, but the performance of the cloud server has not been considered in the literature. It is notable if cloud services are not efficient for offloaded tasks the primary objective of mobile cloud architecture quit meaningless. Existing research did not consider the performance of the cloud resources and transmission delays together which could be degrading the performance of the application. However, current literature research on mobile offloading system needed to be addressed following questions issues:

- **Offloading decision:** it is the fundamental phase in application partitioning what to, how to and where to migrate the application tasks for execution. But there is a big challenge, how to partition the application in an accurate way, in a more general way which factors should be considered for the application partition so that application can be executed within a given deadline.
- **Task Assignment:** mobile cloud application tasks are divided into two disjoint sets, for example local disjoint set of tasks and cloud disjoint set of tasks. It is a challenging how to schedule the tasks on mobile device and cloud resources so that average response time of the application would be minimized without any violation.
- **Environment Adaptation:** however, mobile cloud computing architecture is the amalgamation of different technologies so it can adaptively change the network connection environment (i.e., WIFI, cellular network) when user mobility is facilitated, due to heavy load cloud service resources could be overloaded. To cope with the aforementioned case application portioning and offloading decision must be dynamic and support any adaptation in environment.

To cope up with the above questions we have following contribution answers in this paper:

- For offloading system we are formulating min-cut cost optimization problem. In which application can be partitioned into a local disjoint set and cloud disjoint via min-cut algorithm. To cope up with the min-cut problem with environment adaptation, we have proposed a Dynamic Application Partitioning Task Scheduling which has to be followed phases: (i) partition the application into local execution and cloud execution based on following factors, for example, task size, mobile CPU speed, available network bandwidth, Cloud server speed, and cloud resource availability, (ii) schedule the local tasks on the mobile device (e.g., heterogeneous CPU cores), (iii) schedule cloud tasks on the efficient network wireless channel band and (iv) schedule all offloaded task on the cloud resources. It is noteworthy, mobile scheduler only schedule the local tasks on the mobile device and wireless network where it cannot schedule task on cloud resources, however it can be anticipated the execution time of task on cloud resources.

- Task Assignment: we formulate task assignment mixed as an integer linear programming and scheduling problem. The application deadline is a positive integer. The application execution must be feasible and less than a given positive number. The task scheduling is an efficient to execute the tasks on the mobile device and a cloud resource in an optimal way. The proposed algorithm DAPTS which first order the task sequence and then schedule them on appropriate mobile and cloud resource so that minimize the average response of the application.
- The DAPTS supports environment adaptation when network bandwidth and cloud resources are not sufficient for execution while application in the progress it will re-partitioned the application according to new available parameters.

The rest of the paper is organized as follows. Related works are reviewed in Sect. 2. Sections 3 and 4 describes and formalizes the problem under study. A heuristic is proposed for the considered problem in Sects. 5 and 6. Section 5 evaluates the performance of the proposal under different workload scenarios followed by conclusions in Sect. 6.

2 Related Work

In mobile Cloud Architecture, application partitioning is hot favorite topic nowadays. The Offloading technique is an efficient to allow executing compute intensive applications inside mobile device. Many efforts have been made on application partitioning framework such as in [4] Mobile Assistance User Interface (MAUI) proposed the framework in which the objective was to minimize the device energy during computation offloading. MAUI framework is consisted two run time environments (such as mobile run time and cloud run time) to support mobile cloud application. MAUI has considered only mobile end run time performance it did not consider cloud end offloaded cost. CloneCloud run time environment framework for computational offloading proposed in [5]. The objective is to minimize the device energy consumption. It is highly dynamic technique for computational offloading in which mobile application services wrapped into a virtual machine. Application run time manager captured the full image of the application and offload to the cloud server for execution. However, mobile application and cloud server both have virtual environment for the offloading and execution. CloneCloud run time focused on mobile device execution time performance and did not consider the cloud server end.

Nevertheless, computational offloading to remote cloud is not favorable environment for real time applications because they have required lower end to end latency. On the other hand remote cloud located the multiple hops away from mobile device application, it incurred longer delay. In [6] run time environment for computational offloading to the local cloudlet server has been proposed. Whereas, the local cloudlet is the subset of remote cloud, it enables and extends the remote cloud service at the edge of the network. In order to support the real time application according to given latency bound. Furthermore Cuckoo [7] and Jade [8] run time frameworks have considered only mobile end performance to minimize the latency, minimize the device energy consumption and increase the throughput of the application. In addition, mobile run time frameworks with various objectives has been invested in [9].

With the best of our knowledge, application partitioning and task assignment problem by considering mobile performance and cloud resource performance together has not been studied yet for delay sensitive application. Summary, in this paper we are formulate the application partition and the task assignment problem for delay sensitive application in mobile cloud environment where application run time is executed locally and remotely on the cloud resources, while the application is bounded by deadline constraint. The objective function is to minimize average response time of the application.

3 Proposed Description

We are analyzing an application partitioning and the task assignment problem for delay sensitive application. To cope up with the problem we have proposed a novel mobile cloud offloading architecture as shown in Fig. 1. Mobile end user submitted healthcare application (Health-APP) as a thin client. It starts with static analysis technology which tells about native and remote task which is developed during design time. However, after application install by mobile user mobile master node performs some operation via following components: (i) task management component is the responsible to estimate the task execution time on the mobile side and on the cloud resources in advance and sequence the tasks in specified order, (ii) offloading decision makes application partitioning decision based on available bandwidth, mobile CPU, server speed, (iii) partitioning results component shows which tasks are executed locally and which tasks are offloaded to the server, then mobile scheduler send offloaded tasks to the cloud via wireless channel band and remaining entire tasks are schedule locally on the mobile device. On the other cloud scheduler allocate the resources to the offloaded tasks and sent back their results to the mobile device after execution.

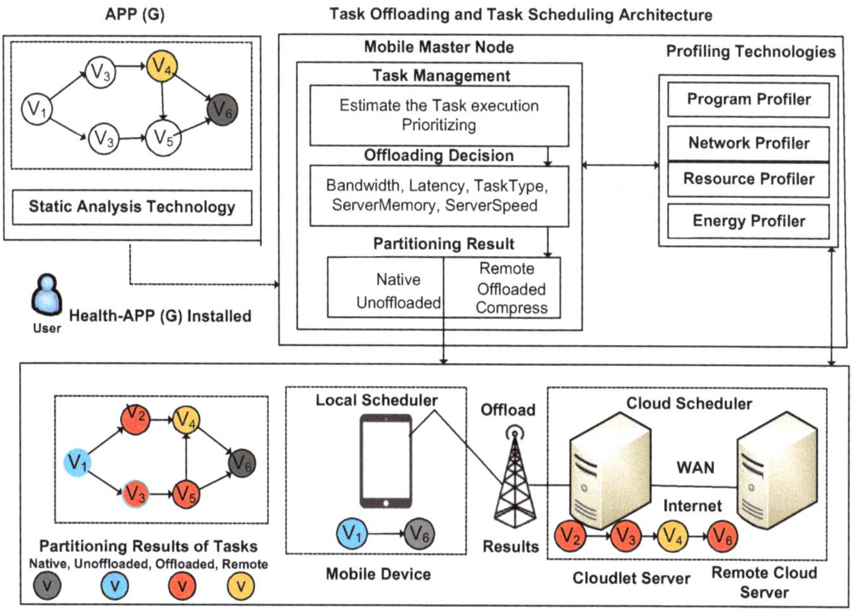

Fig. 1 Mobile cloud application partitioning and task assignment

3.1 Proposed Model and Problem Formulation

3.2 Task Execution Scenario

We have made following assumption to task execution: (i) Each application v_i can be executed on a mobile device or on a cloud resource, it depends on task characterization such as a task data size, if a task is compute intensive it need to be offloaded to the cloud server else execute on a local mobile, (ii) network connection is not fixed between mobile user application and cloud server for both uploading and downloading data, (iii) offload task must be compressed with fixed ratio, in general terms the offloaded tasks size must 6–10 MB, (iv) application response time has linear relationship with task execution time.

3.3 Application Characteristics

A Mobile Cloud Architecture (MCA) is the combination of the wireless network, mobile edge server and cloud server. The Mathematical Notations are marked in Table 1. In MCA a workflow, mobile healthcare application is modeled as consumption weighted Directed Acyclic Graph (DAG) i.e., G (V, E). Whereas, each task v_i is represented by a node $v_i \in V$. An edge $e(v_i, v_j) \in E$ represents

Table 1 Mathematical notation

Notation	Description
V	Number of tasks v
$\mathcal{V}_Z$	Number of virtual machines $\mathcal{V}$
$\mathcal{V}_j$	jth virtual machine in edge server
v_i	Workflow application task
W_i	Weight of the each task
D_G	Deadline of the application G
ζ_j	Speed of jth virtual machine
ζ_m	Speed of mobile processor v_i
C_i^e	Execution cost of task v_i on cloud
x_{ij}	Assignment of task v_i on virtual machine j
B_i	Begin time of the task v_i
F_i	Finish time of the task v_i
G	DAG graph
A	Most Tightly Connected Vertices
a	Arbitrary of vertex G

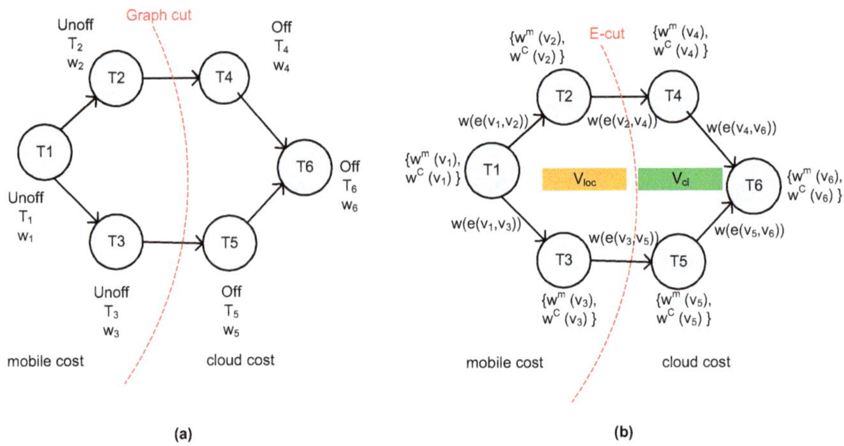

Fig. 2 Application partitioning consumption graph. (**a**) Application consumption call graph. (**b**) Application weighted consumption graph

communication between v_i to v_j. A task v_i could be started until associated all predecessors complete. v_1 and v_n are two dummy tasks (i.e., entry task and exist task). A task could be started until associated predecessors complete. In simple words v_j cannot be started in anticipation of v_i get the job done and $i < j$. We partition the call graph into consumption graph as shown in Fig. 2b based on static analysis technology and profiling technology, we convert the call graph into a task weighted consumption graph after offloading decision as explained in architecture. We formulate this problem as min-cut cost problem and we divide the mobile application tasks the local disjoint set, i.e., V_{loc} and V_{cl} by min-cut. The application is bounded by deadline constraint D_G.

3.4 Mobile Task Assignment

Each task has w_i as input data, and it generates the output data w_i, it requires mobile CPU m instruction for execution (per second), where mobile device has M CPU cores, i.e., $M = \{m_1, m_2, .., M\}$ with homogeneous speed. The decision variable shows that either task executes on mobile core or not and denoted as a $y = \{v_i \in V, m \in M\}$, whereas, yi, m={0, 1}. The $yi, m = 1$ or $yi, m = 0$ shows assignment either an execution occur on mobile core or not. The execution time of the each task on the mobile device can be calculated in the following way:

$$T_i^{loc} = \frac{w_i}{\zeta_m}, \tag{1}$$

the average execution of all tasks on local mobile device can be expressed in the following way:

$$\sum_{i=1}^{V} \sum_{m=1}^{M} T_i^{loc} \times y_{i,m}, \tag{2}$$

mobile scheduler can scheduler one task at a time. However, mobile device can schedule remote task on the wireless channel band one task a time, offload task incurs with communication time which could be expressed in the following way:

$$T_{i,j}^{e} = \left(w \frac{(i, j}{B_{upload}}\right) + \left(w' \frac{(j, i}{B_{download}}\right), \tag{3}$$

whereas, average communication cost can be calculated explained in the following way:

$$\sum_{e(v_i, v_j) \in E} T_{i,j}^{e}. \tag{4}$$

3.5 Cloud Task Assignment

A set of cloud edge servers can be represented by K={k_1, ..., k_n}. We presuppose that each k server holds one virtual machine type, that all virtual machine instances are heterogeneous, every virtual machine (VM) has dissimilar computation speed which are illustrated as $\zeta_j = (j = 1, \ldots., Z)$. A set of virtual machine instance can be shown by VK={vk_1, ..., vk_n}, in which K_i^{vk} is the virtual machine assignment for v_i. Each workflow application has workload W_i={$i = 1$, ..., N} with deadline D_i. To minimize the response time of the submitted workflow tasks, we assign each application task to the optimal VM while meeting the deadline D_i, because the optimal VMs always leads to lower response time. Since a task v_i is only can be performed by one VM j, a decision variable $x_{ij} \in \{0, 1\}$ is utilized, x_{ij}=1 only if the task v_i is assigned to the VM V_j.

$$T_i^{cl} = \frac{w_i}{\zeta_j}, \tag{5}$$

the average execution of all tasks on cloud virtual can be expressed in the following way:

$$\sum_{i=1}^{V} \sum_{j=1}^{Z} T_i^{cl} x_{i,j} \times T_i^{cl}. \tag{6}$$

In the same way, the vector $Pr_v = \{V_{loc} \in V, V_{cl} \in V\}$ with variable indicates whether execution offloading or not, namely:

$$Pr_v = \begin{cases} 1, \text{ if } v_i \in V_{loc} \\ 0, \text{ if } v_i \in V_{cl} \end{cases} \tag{7}$$

According to Eq. (4), the communication cost is determined by E_{cut}=1, otherwise same location tasks have no communication cost as shown in Fig. 2.

$$Pr_e = \begin{cases} 1, \text{ if } e \in E_{cut} \\ 0, \text{ if } e \notin E_{cut} \end{cases} \tag{8}$$

3.6 *Application Response Time*

The total response time of workflow application is an amalgamation of computation time and communication time. Since, computation cost could location and remote execution after application partitioning. The communication cost is determined by weight of data transport and available network bandwidth. The average response of workflow application due to offloading is expressed as follows:

$$T_{total} = \sum_{v \in V} Pr_v.T_v^{loc} + \sum_{v \in V} (1 - Pr_v).T_v^{cl} + \sum_{e(v_i, v_j) \in E} Pr_e.T_e^{trans}, \tag{9}$$

whereas, Eq. (9) describes that the average response of workflow application is the sum of local and remote computation cost and communication cost. The considered problem is mathematically modeled as bellow:

$$\begin{aligned} &\underset{T}{\text{minimize}} \quad T_{total} \\ &\text{subject to} \quad T_{total} \leq D_G, \end{aligned} \tag{10}$$

$$T_{j,0} = 0, T_{m,0} = 0, \tag{11}$$

$$\sum_{i=1}^{V} \sum_{m=1}^{M} T_i^{loc} \times y_{i,m}, \sum_{i=1}^{V} \sum_{j=1}^{Z} T_i^{cl} x_{i,j} \times T_i^{cl}, \tag{12}$$

$$F_i^{cl} = \sum_{j=1}^{Z} T_{i,j}^{cl} x_{i,j}, F_i^{loc} = \sum_{m=1}^{M} T_{i,m}^{loc} y_{i,m}, \tag{13}$$

$$\sum_{i=1}^{V} x_{i,j} = 1, \sum_{j=1}^{Z} x_{i,j} = 1, \sum_{m=1}^{M} x_{i,j} = 1, \tag{14}$$

$$T_{total} \leq D_G, x_{i,j}\{0, 1\}, \tag{15}$$

$$x_{i,j}\{0, 1\}, \tag{16}$$

$$y_{i,m}\{0, 1\}, \tag{17}$$

However, Eq. (9) calculates the average response time of all tasks. In this paper, we assume at the start and end of the task on a cloud virtual machine, because all compute intensive tasks are executed on the cloud. According to Eq. (10) initial of any virtual machine is assumed to be zero. The finish time of task v_i on virtual machine j and mobile device m is $T_{i,j}$, $T_{i,m}$ determined by the previous task v_i-1 execution time that is $\sum_{v=1}^{V} x_{i,j} T_i^{cl}$, $\sum_{v=1}^{V} y_{i,m} T_i^{loc}$. Equation (11) determines the execution cost of a task on the cloud and mobile device according to their speed and weight. The task finish time is determined by Eq. (12). Equations (13), (15), (16) demonstrates that each task can only be assigned to one virtual machine and the each virtual machine can be only assigned to on Task. According to Eq. (14) finish time of application G should be less than a give deadline D_G.

4 Proposed Algorithm

The objective function is to minimize the average response time (i.e., total time) of application which has a hard constraint deadline. Each application should follow the precedence constraint rule, its general terms one task execute at a time either on a mobile device or offloaded to the cloud server for execution. For considered problem we have proposed, a novel DAPTS algorithm which has following phases:

(i) it starts with application partitioning based on static analysis technology and profiling technology, (ii) offloading decision is an important phase in which we divide the application into local execution and remote cloud based on following factors: network bandwidth, task size, task type (unoffload, offloaded), cloud server speed, mobile CPU speed and server memory., (iii) after partitioning resulting mobile scheduler schedule all native tasks on the mobile cores and schedule remote tasks on the efficient wireless channel band, (iv) cloud schedule all offloaded tasks and after execution their results sent back to the mobile device. However, for phase (i) and (ii) we formulated min-cut problem and proposed Algorithms 1, 2 and 3 based on min-cut traditional algorithm. For phase (iii) and (iv) we formulate task scheduling problem, and proposed iterative Algorithm 5 which effectively execute the all tasks without violate any application deadline. The Algorithm 1 shows we can optimize phase (i) and (ii) based on min-cut algorithm, whereas, phase (iii) and (iv) can be optimize based on iterative algorithm 5.

Algorithm 1 DAPTS framework

```
Input  : v_i ∈ V ;
Output: min_Z;
begin
    Z ←0;
    Application Partitioning based on Eq. (2) Workflow Task Sequence;
    foreach v_i ∈ V do
        Optimal VM Searching ;
    Task Sequence Adjustment;
    return Z;
```

4.1 Application Partitioning Phase (i) and (ii)

For phase (i) and (ii) the application partition algorithm is formulated as the min-cut procedure (i.e., function) in algorithmic 2, whereas every phase j it called the procedure Min-Cut-Phase function while explained in algorithmic 3. While few tasks will have to be either executed locally or remotely, we need to be merged all of them into new single node. The basic fundamental of this algorithm to make the process very easy to select the subsequently vertex that would be added to the given vertex set A, that is MTCV (Merge Most Tightly Connected Vertex) as shown in Fig. 3, which is described as the node whose $\Delta(v)$ to vertex A is highest. Nevertheless, $\Delta(v)$=$w(e(A,v))$-[$w^{loc}(v) - w^{cl}(v)$]. Additional, the entirety cost of partitioning is followed:

$$Cost_{cut(A-t,t} = C^l oc - [w^l oc(v) - w^{cl}(v)] + \sum_{v \in A \setminus t} w(e(A,v)). \quad (18)$$

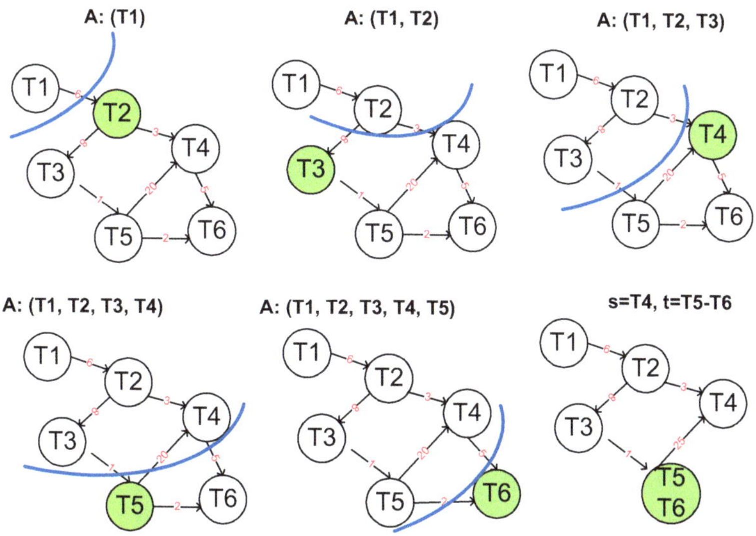

Fig. 3 Merging tightly connected vertex to A

Let assumed the local cost at the mobile device $C^m = \sum_{v \in V}[w^{loc}(v)]$ and same formula for remote cost, while the cut-value $Cost_{cut(A-t,t)}$ is the partitioning cost, $w^{loc}(g) - w^{cl}(g)$ is the gain g vertex from offloading and $\sum_{v \in A \setminus t} w(e(A, v))$ is communication cost incurred by offloading. Input is $G(V, E)$ workflow application edges with non-negative incident weight, output return minimum cut of G as shown in Fig. 3.

Algorithm 2 Merging function

Input: (G,w);
begin
  $g_{s,t} \Leftarrow s \cup t$;
  for *graph all vertices* $v \in V$ **do**
    if $v \neq \{s, t\}$ **then**
      $w_t(e(g_{s,t}, v)) = w_t(e(s, v)) + w_t(e(t, v))$;
      // all edges weights added here
      $[w^m(g_{st})], [w^{rc}(g_{st})] = [w^m(s)] + [w^m(t)], [w^{rc}(s)] + [w^{rc}(t)]$;
      // all nodes weights added here
      $E \Leftarrow E \cup e(g_{st}, v)$;
      // edges are added
    $E^* \Leftarrow E \setminus \{e(s, v)\}\{e(t, v)\}$;
  $V^* \Leftarrow E \setminus \{s, t\} \cup g_{s,t}$;
  return $G^*(V^*, E^*)$

Algorithm 2, merge those all vertices which are closely related to MTVC A. Algorithm 4 always return optimal min-cut as shown in Fig. 4.

Algorithm 3 Min-cut function

Input : (G, w);

begin

  $w_t(min_{cut} \Leftarrow \infty$;

  for *m=1;length(source-vertices)* **do**

    `// Off and Unoff vertices merged into single node`

    (G, w_t)= merging(G, w_t, source-vertices, source-vertices(m));

    while $| v \in V | > 1$ **do**

      $[C_{ut}(A - t, t), s, t]$=Min-Cut-Phase(G, w_t);

      if $w_t(c_{ut}(A - t, t)) < w_t(min_{cut})$ **then**

        $min_{cut} \Leftarrow c_{ut}(A - t, t)$;

    merging(G, w_t, s, t);

  return min_{cut}, Min-Cut-Grouping-List;

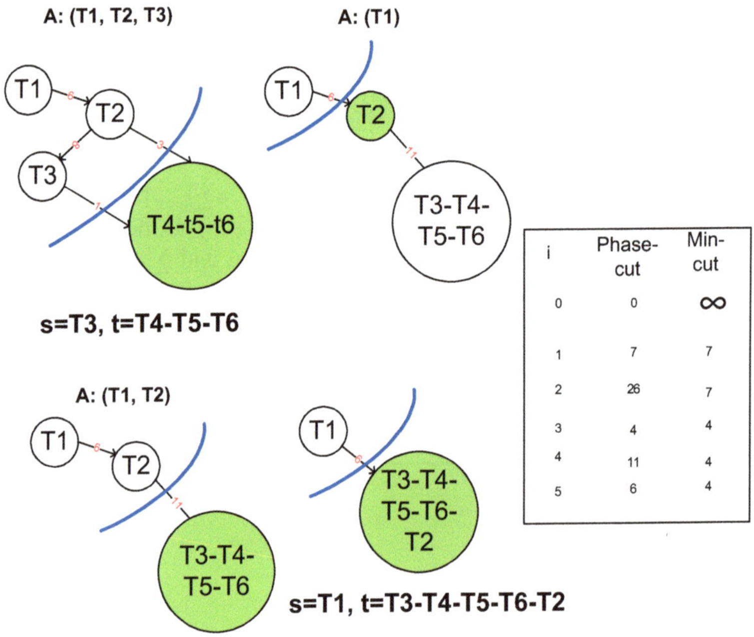

Fig. 4 Application partitioning consumption graph

4.2 *Task Assignment Phase (iii) and (iv)*

The aim of the task scheduling problem is to find a task assignment and the start times of the tasks to the processors in a way that the schedule span is minimized and, in the same time, the precedence constraints are preserved. For phase (iii) and (iv) we have proposed greedy algorithm based iterative Algorithm 5, in which we can execute the application tasks on appropriate resources. We perform here static scheduling mechanism, whereas proposed algorithm 5 is involved in three stages:

first we sequence the task based on HEFT algorithm [10], second stage, sort out all the mobile cores and virtual machines in optimal order, third stage, schedule all tasks to the optimal appropriate mobile cores and cloud virtual machine so that minimize maximum lateness and execute them within application deadline. The task sequence ordering in the following way:

- Shortest deadline First (SDF): We sort the set of workflow applications based on their deadline. The small deadline application sort first and the bigger one later. If the deadline is same then FCFS policy will be apply.
- Shortest slack time first (STF): The application tasks are sorts according to the task slack time (TST). The task which has shortest slack time, they schedule first.
- Shortest weight First (SWF): The applications are sequenced based on the weight of all tasks, shortest weight application arranged first and the bigger one later.

Algorithm 4 Min-cut-phase function

```
Input : (G, w);
begin
    a ⇐ random node of G_j;
    Added vertex a ⇐ {A};
    while [A ≠ V_j] do
        Maximum=-∞;
        v_maximum=null;
        for v ∈ V_j do
            if v ≠ A then
                // Performance add a task v to the remote cloud
                Δv ⇐ w_t(e(A, v)) − [w^m(v) − w^rc(v)] // Determine the node
                    that is the MTCV to A
                if Maximum≤ Δv then
                    Maximum=Δv;
                    v_maximum = v;
        A ⇐ A ∪ {v_maximum};
        a ⇐ merging(G_j, w_t, a, v_maximum);
    Lastly vertex source s would be added to the MTCV A;
    Lastly vertex sink t would be added to the MTCV A;
    return min_cut(A − t, t), ;
```

For sequence the task slack time T_i^{slack} must be minimized. The slack time T_i^{slack} of the task v_i is found by the definite finish time F_i and the deadline D_G i.e., $T_i^{slack}=F_i-D_G$. The finish time F_i of the task v_i is determined by the execution time T_i^{cl}, T_i^{loc} and the available time of the assigned mobile core and virtual machine. Nevertheless, T_i^{cl}, T_i^{loc} must be determined earlier than scheduling, the regular execution time T_i^{slack} is mathematically defined as:

$$T_i^{slack} = D_i - F_i \tag{19}$$

$$T_i^{slack} = D_i - F_i \tag{20}$$

$$F_i = \sum T_i^e \frac{\sum_{j=1}^{M} W_i}{\sum_{j=1}^{M} \zeta_i} \tag{21}$$

4.3 Greedy Task Scheduling Algorithm

The task sequence is produced by the proposed task sequencing method; the goal is to assign tasks to optimal mobile core and cloud virtual machines, which has lower lateness. The existing approach The allocation of each task v_i to a mobile core and virtual machine is to make a decision on the $x_{i,j}$, the $x_{i,j} = 1$ if the v_i is assigned to the VM $\mathcal{V}_j$ As depicted in Eq. (21), the optimal virtual machine is decided by the ζ_j speed of the assigned VM and the task execution time T_i^e. The optimal virtual machine selection is defined as follows:

$$\zeta_j^* = \frac{w_i}{\zeta_j} \tag{22}$$

Optimal mobile core for local task assignment is expressed in the following way:

$$\zeta_m^* = \frac{w_i}{\zeta_m} \tag{23}$$

The greedy Algorithm 5 performs on phase (iii) and (iv), line 2–3 sort out all mobile cores and the virtual machine in ascending order in the context of their speed. Line 4–13 calculate the execution on all possible cloud virtual machines with minimum lateness and calculate the average execution of all tasks on the cloud. Line 15–23 the execution on all possible mobile cores with minimum lateness and calculate the average execution of all tasks on the mobile cores. Line 24–25 describe that the total execution of all tasks must be less than the given deadline. Line 26 returns optimal average response time of the application after optimal searching appropriate virtual machines and mobile cores.

4.4 Time Complexity

The running time of DAPTS is $O(N \times N^3)$=n^4. The DAPTS algorithm is iterative in nature. Previously proposed framework [4–6] and [11] have running time more than n^5. Proposed algorithm DAPTS works better even though in worst case. For simulation, we have application scenario based on Fig. 5, the scenario is the combination of the mobile computing resources, network base stations and cloud resources (edge cloud servers and remote cloud server). Each application tasks can be executed either on locally on the mobile cores or remote cloud. The fundamental time is the computation time (i.e., local execution time and cloud execution time) and communication time when remote tasks offloaded to a base station or wireless channel band to the cloud server for the computation.

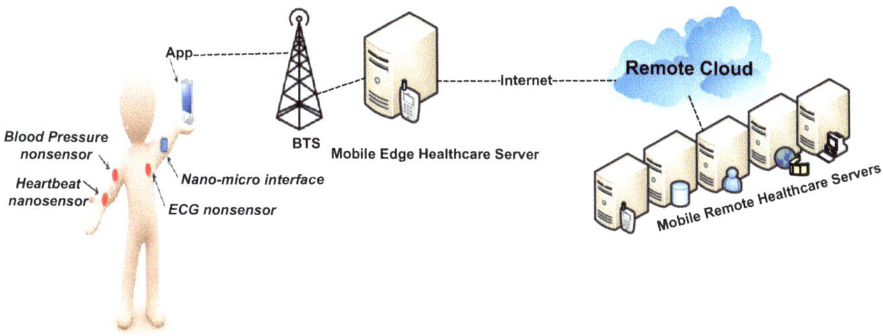

Fig. 5 Application simulation scenario

Algorithm 5 Optimal mobile and VM searching

Input : $(v_i \in V)$ to Scheduling;
Output: $\min_T$;
begin
  $Q_{vm} \leftarrow$ Sort all VMs by their speed ζ_j^* with ascending order;
  $Q_m \leftarrow$ Sort all mobile by their speed ζ_m^* with ascending order;
  $\mathcal{V} \leftarrow$Null;
  $m \leftarrow$Null;
  foreach $\mathcal{V}_j \in Q_{vm}$ **do**
    $T_{j,0} \leftarrow 0$;
  foreach $\mathcal{V}_j \in Q_{vm}$ **do**
    Calculate T_i^{cl} of $\mathcal{V}_|$ based on Eq. (8);
    if $T_{j,i-1} + T_i^{cl} = 1$ **then**
      Calculate the $T_{j,i}$ of $\mathcal{V}_j$ by Eq. (7);
      $\mathcal{V} \leftarrow \mathcal{V}_j$;
      break;
    Calculate the average execution of all tasks based on Eq. (6);
  foreach $\mathcal{V}_j \in Q_{vm}$ **do**
    $T_{j,0} \leftarrow 0$;
  foreach $m \in Q_m$ **do**
    Calculate T_i^{loc} of m based on Eq. (11);
    if $T_{m,i-1} + T_i^{loc} = 1$ **then**
      Calculate the $T_{m,i}$ of m by Eq. (12);
      $m \leftarrow m$;
      break;
    Calculate the average execution of local tasks based on Eq. (2);
  if $m + \mathcal{V} \leq D_G$ **then**
    $T_{total} = m + \mathcal{V}$;
  return T

5 Performance Evaluation

In this paper, we have divided the evaluation part into two phases such as optimal application partitioning phase and task scheduling phase.

5.1 *Application Evaluation*

To estimate the partitioning DAPTS algorithm, the following fixed values must be known in advance. For example, **fixed values:** these values are closely related to power consumption P_m, P_i, and P_{tr} and to the specific mobile device. The configuration we have employed such as PDA HP IPAQ with large Intel-Scale processor by subsequent values: $P_{tr} \approx 1.3W$, $P_m \approx 0.3W$ and $P_i \approx 0.7W$. **Precise values:** these parameters are closely related data transfer size, network upload and download and current mobile and cloud speed factor F. **Fluctuation values:** these values are scrupulously to mobile device current workload status, network status and cloud status. Due to adaption and fluctuation, it is not trivial to calculate this value initially. The profiling (program (i.e., mobile workload and other status) and network (i.e., bandwidth and available 3G/4G, WIFI and etc) can be enabled to calculate these values during fluctuation and adaption.

5.2 *Task Scheduling Evaluation*

In this paper, proposed algorithm DAPTS is composed of various parameters and components. However, the calibration algorithm parameters and components, the evaluation relating the proposed DAPTS and huge benchmark heuristics [12–14] which are already employed in the WorkflowSim Cloudsim stand. However, existing workflowsim only hold ups, remote cloud data center services, the platform services extended to the proximity mobile edge closer to the mobile user network in this paper. The simulation parameters are explained in Table 2.

5.3 *Simulation Setup*

Simulation setup again divides the application workload into mobile execution and remote execution, and then schedule related tasks along their respective processor.

Dynamic Application Partitioning Setup The healthcare workflow application is partitioned as depicted in Fig. 6. Each task is represented by a node, whereas, each node has exactly two costs (i.e., local execution and cloud execution cost). We partitioned the application under F=2 speed up factor and available bandwidth =1

Table 2 Simulation parameters

Simulation parameters	Values
λ_i user arrival time	5 s
Languages	JAVA, XML, Python
Applications	A.G, Healthcare
Application workload	Tasks
Application fine-grained	Methods
No. of tasks	1/250
Benchmark workload	Augmented Reality (AG)
Simulation time	6 h
Experiment repetition	14
No. of mobile devices	100–1000
Location user mobility	M-M-Nomodic
WAN-WLAN Network Bandwidth	20–300 mbps
WAN-Propagation Delay	50–150 mbps
Standard task size	1500–2000 MI
Upload/download data size	2000/150 KB
Possibility offload to edge cloudlet	80%
Possibility offload to remote cloud	12%
Container processing speed cloudlet/cloud	1200/22,000 MIPS
No. VMs per cloudlet/cloud	3/∞
No. of Mobile heterogeneous cors	6
No. of VMs.	10–200
Container speed	500–2500 MIPS
Container RAM	2–4 GB
CPU-Utilization for Healthcare APP. cloudlet/cloud	15–0

M/B respectively. Still, blue nodes are performed locally, and reds are offloaded to the cloud for performing. However, DAPTS will re-partition the application if the wireless bandwidth B or the speedup factor F diverges.

Task Scheduling Setup Task scheduling is not trivial with optimal assignment [15]. For task scheduling, algorithm values should calibrate the components and parameters, and workflow applications are generated randomly [16]. Healthcare workflow applications are created with different five sizes such as $Q_w \in \{20, 40, 60, 80, 100\}$. Since, each healthcare workflow application is comprised of four unlike figures of tasks i.e., $Q_t \in \{50, 100, 200, 500\}$. We produce ten combinations of Q_w and Q_t respectively. However, each healthcare workflow application is bounded by deadline. The deadline of each workflow D_w is expressed as follows:

$$D_w = T_w^{f_i} + \gamma + T_w^{f_i}, \tag{24}$$

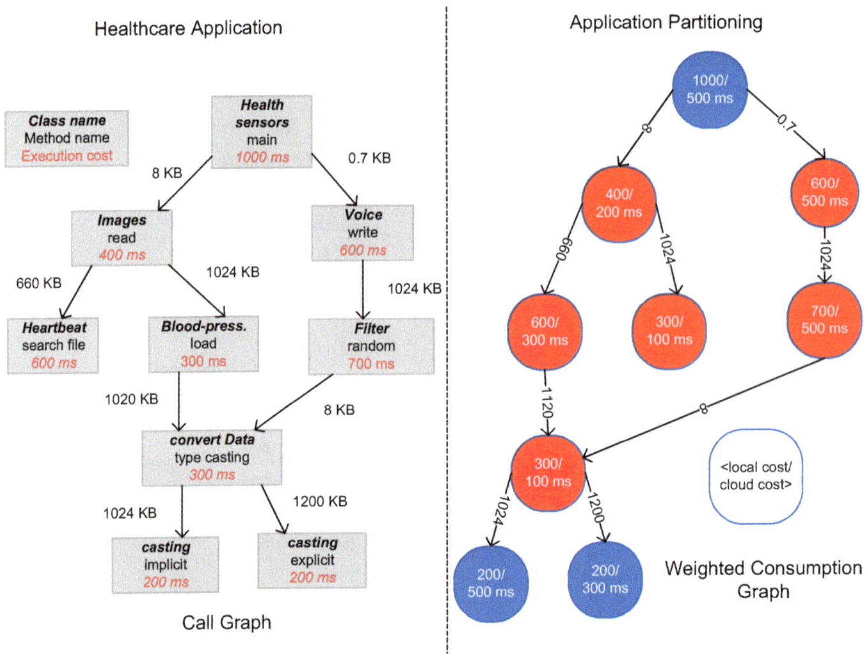

Fig. 6 Healthcare application partitioning

$T_w^{f_i}$ is the workflow of tasks with finish time, whereas f_i is calculated based on Eq. (15). A γ is described as a factor to control the tightness of the deadline D_w, and $\gamma \in \{0.2, 0.4, 0.6, 0.8, 1\}$. Hence, each healthcare workflow application exactly has diverse deadline i.e., $\{D_1, D_2, D_3, D_4, D_5\}$. To evaluate the performance of the proposed DAPTS algorithm next to healthcare heuristics based DEA benchmark [17] and [18] to verify the strength of the proposed algorithm. The calibration parameters of tasks are same as a healthcare workflow, the performance evaluation of the DAPTS is measured at diverse deadlines since strict to lose. The DAPTS has RPD (Relative-Percentage-Division) is utilized to compare with existing schemes such as non-offloading and described as follow:

$$RPD\% = \frac{Z - Z^*}{Z} \times 100\%, \tag{25}$$

Whereas, Z is the local cost i.e., T_{loc}, and Z^* is the total cost $T_{total} = \sum_{v \in V} Pr_v.T_v^{loc} + \sum_{v \in V}(1 - Pr_v).T_v^{cl} + \sum_{e(v_i,v_j) \in E} Pr_e.T_e^{trans}$. The evaluation is carried out in mobile cloud architecture, which is geo-graphically distributed in nature. In MCA has following virtual machine configuration: VM_1 (Core: 1, MIPS: 200, RAM 1024 GB RAM), VM_2 (Core:1, MIPS: 400, RAM 2048 GB RAM), VM_3 (Core:1, MIPS: 600, RAM 3072 GB RAM), VM_4(Core:1, MIPS: 800, RAM 4096 GB RAM), the CPU of every VM is locate to be 1, but the processing speed are diverse. Each type of VM is generated randomly. Each created VM must be matched and optimal earlier than calibrating parameters and components towards algorithm evaluation.

5.4 Parameter and Components Calibration

In the DAPTS framework, there are six components for the healthcare workflow application i.e., partitioning and sequencing: Merging function, min-cut phase and min-cut, SDF, ST, and SWF respectively. Each workflow application tasks is represented by n_w, and $\tau \in \{0.2, 0.4, 0.6, 0.8, 1\}$ is tightly value of workflow application G_w.

5.5 Algorithm Comparison

Based on the ANOVA technique, Fig. 7 describes that the mean plot of τ by 95.0% Tukey HSD intervals. It can observe that RPD% reduces when τ increases from 0.2 to 0.4. Existing heuristic such as full offloading (FUL), non-offloading (NOF), partial offloading (PAR) is compared with proposed DAPTS based algorithm components. As we described above in DAPTS has six components likewise, Merging function, min-cut phase and min-cut, SDF, ST, and SWF respectively. We evaluated the overall performance and validity based on above components, and all RPDs results proposed algorithm are better as compared to all benchmark heuristics bounded by deadline constraints [19] and [20]. According to Fig. 8 proposed algorithm RPD% is better on different speedup factor F and B bandwidth than existing heuristic techniques. Figure 9 shows that healthcare workflow application tasks are completed within a given deadline after partitioning. The RPD% of proposed algorithm DAPTS is optimal in all workflow applications as shown in Fig. 10.

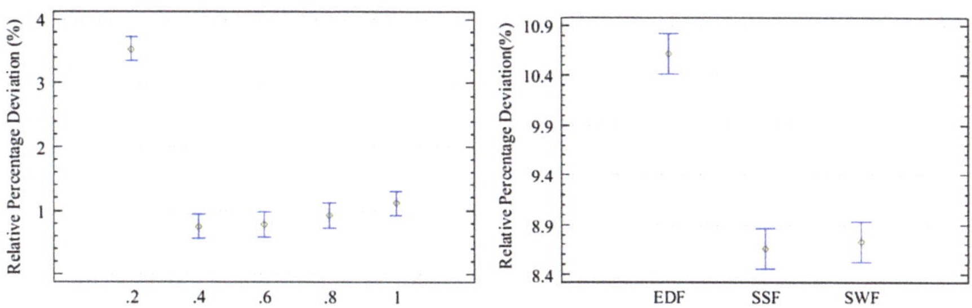

Fig. 7 The mean plot of with 95.0% Tukey HSD intervals and The mean plot of workflow sequence rules with 95.0% Tukey HSD intervals

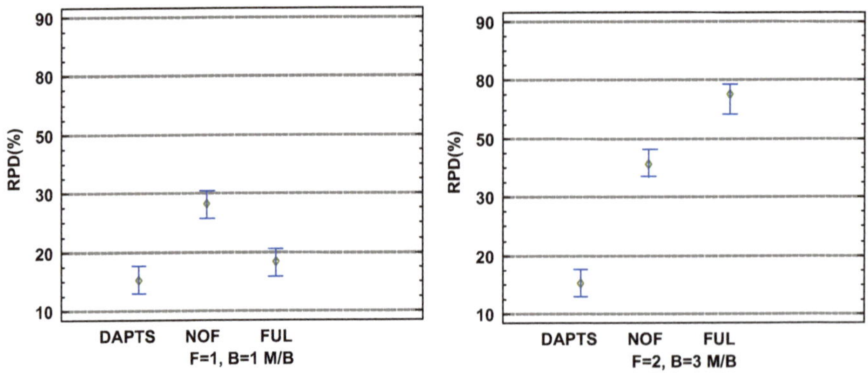

Fig. 8 Application partitioning performance based on speed up factor F and bandwidth B

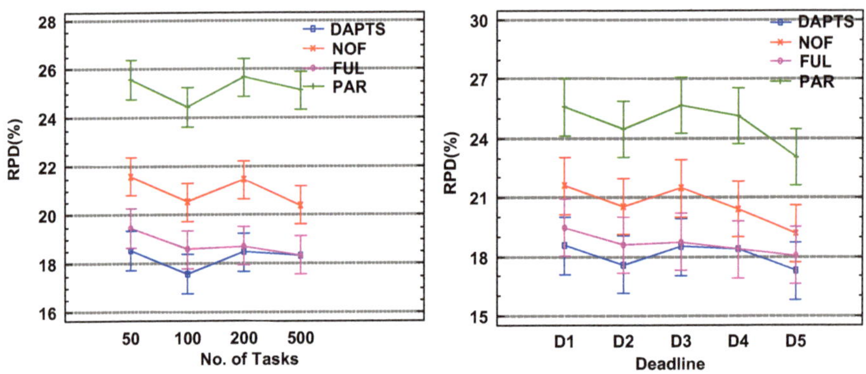

Fig. 9 Workflow application no. of tasks completed within a given deadline

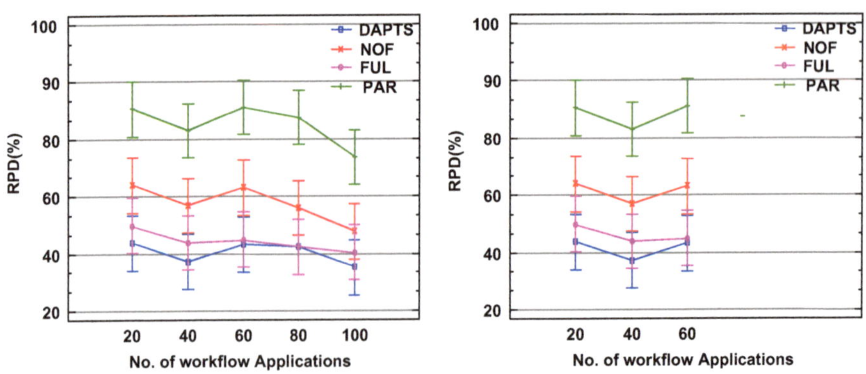

Fig. 10 Multiple workflow applications

6 Conclusion

In this paper, we have proposed dynamic application partitioning (DAPTS) algorithm and task scheduling for real time healthcare application. We have evaluated our proposed algorithm in the healthcare benchmark workflow application and comparison with benchmark heuristics, and finish all workflow tasks within a given deadline. In this paper, our goal is to minimize the average response time of all healthcare applications. Another hand, optimal the virtual machine is allocated to workflow application. In future work, we will study bi-objective such response time and energy consumption minimization in the dynamic application partitioning problem.

References

1. Mahesar, A.R., Lakhan, A., Sajnani, D.K., Jamali, I.A.: Hybrid delay optimization and workload assignment in mobile edge cloud networks. Open Access Lib. J. **5**(09), 1 (2018)
2. Aamir, M., Hong, X., Tahir, M., Wagan, A.A.: Cloud compting and associated mitigation techniques: a security perspective. J. Emerg. Trends Comput. Inf. Sci. **5**(3), 165–171 (2014)
3. Lee, H.-S., Lee, J.-W.: Task offloading in heterogeneous mobile cloud computing: modeling, analysis, and cloudlet deployment. IEEE Access **6**, 14908–14925 (2018)
4. Shiraz, M., Gani, A., Khokhar, R.H., Buyya, R.: A review on distributed application processing frameworks in smart mobile devices for mobile cloud computing. IEEE Commun. Surv. Tutorials **15**(3), 1294–1313 (2013)
5. Chun, B.-G., Ihm, S., Maniatis, P., Naik, M., Patti, A.: Clonecloud: elastic execution between mobile device and cloud. In: Proceedings of the sixth conference on Computer systems, pp. 301–314. ACM, New York (2011)
6. El-Barbary, A.E.-H.G., El-Sayed, L.A.A., Aly, H.H., El-Derini, M.N.: A cloudlet architecture using mobile devices. In: 2015 IEEE/ACS 12th International Conference of Computer Systems and Applications (AICCSA), pp. 1–8. IEEE, Piscataway (2015)
7. Kemp, R., Palmer, N., Kielmann, T., Bal, H.: Cuckoo: a computation offloading framework for smartphones. In: International Conference on Mobile Computing, Applications, and Services, pp. 59–79. Springer, Berlin (2010)
8. Qian, H., Andresen, D.: Jade: an efficient energy-aware computation offloading system with heterogeneous network interface bonding for ad-hoc networked mobile devices. In: 2014 15th IEEE/ACIS International Conference on Software Engineering, Artificial Intelligence, Networking and Parallel/Distributed Computing (SNPD), pp. 1–8. IEEE, Piscataway (2014)
9. Akherfi, K., Gerndt, M., Harroud, H.: Mobile cloud computing for computation offloading: issues and challenges. Appl. Comput. Inform. **14**(1), 1–16 (2016)
10. Wu, S., Niu, C., Rao, J., Jin, H., Dai, X.: Container-based cloud platform for mobile computation offloading. In: 2017 IEEE International Parallel and Distributed Processing Symposium (IPDPS), pp. 123–132. IEEE, Piscataway (2017)
11. Zhao, Y., Thomas, N.: Performance modelling of optimistic fair exchange. In: On Analytical and Stochastic Modeling Techniques and Applications, 298C313. IEEE, Piscataway (2016)
12. Xu, Y., Zhang, F., Chou, J.: A novel petri nets-based modeling method for the interaction between the sensor and the geographic environment in emerging sensor networks. In: On Analytical and Stochastic Modeling Techniques and Applications, 298C313. ACM, New York (2016)

13. Tsai, C.-W., Huang, W.-C., Chiang, M.-H., Chiang, M.-C., Yang, C.-S.: A hyper-heuristic scheduling algorithm for cloud. In: IEEE International Conference on Cloud Computing Technology and Science, vol. 2(2), pp. 236–250 (2014)
14. Folkerts, E., Alexandrov, A., Sachs, K., Iosup, A., Markl, V., Tosun, C.: Benchmarking in the cloud: what it should, can, and cannot be. In: Technology Conference on Performance Evaluation and Benchmarking, pp. 173–188 (2012)
15. Lin, X., Wang, Y., Xie, Q., Pedram, M.: Energy and performance-aware task scheduling in a mobile cloud computing environment. In: CLOUD '14 Proceedings of the 2014 IEEE International Conference on Cloud Computing, pp. 192–199 (2014)
16. Lakhan, A., Xiaoping, L.: Energy aware dynamic workflow application partitioning and task scheduling in heterogeneous mobile cloud network. In: 2018 International Conference on Cloud Computing, Big Data and Blockchain (ICCBB), pp. 1–8. IEEE (2018)
17. Ettorchi-Tardy, A., Levif, M., Michel, P.: Benchmarking: a method for continuous quality improvement in health. Health Policy **7**(4), e101 (2012)
18. Kane, C.K., Emmons, D.W.: New data on physician practice arrangements: private practice remains strong despite shifts toward hospital employment. Am. Med. Assoc. **4**(6), 1–16 (2013)
19. Lakhan, A., Li, X.: Transient fault aware application partitioning computational offloading algorithm in microservices based mobile cloudlet networks. Computing 1–35 (2019)
20. Kruk, L., Lehoczky, J., Ramanan, K., Shreve, S.E.: Heavy traffic analysis for EDF queues with reneging. Ann. Appl. Probab. **21**(2), 484–545 (2011)

Clustering Priority-Based User-Centric Interference Mitigation Scheme in the Ultra Dense Network

Guomin Wu, Guoping Tan, Fei Feng, Yannan Wang, Hanfu Xun, Qi Wang, and Defu Jiang

1 Introduction

It is common knowledge that UDN is one of the key technologies to promote the development of 5G technology [1, 2]. It increases its density for low-power node deployment, which makes them closer to terminal and greatly improves system capacity. At the same time, spectrum efficiency and power efficiency are improved as well. The UDN can maximize the advantages of proximal transmissions and spatial availability for system resources [3]. However, due to the reduction of node spacing, the transmission loss of adjacent nodes may make more serious interference for terminal [4]. As to the aforementioned problem, how to improve user performance with network cooperation and interference management shows to be critical, which becomes a challenge in UDN research field.

In recent years, researchers have proposed many solutions to deal with inter-cell interference issues [5]. For example, a gossip based distributed power control (GBDPC) algorithm is proposed to combat the co-channel interference in a distributed manner [6]. Give another example, the interference list can be built to construct virtual cell clusters. Then corresponding beamforming matrices can be designed for all user equipment [7]. The performance limits of ultra-dense cloud

G. Wu
Hohai University, Nanjing, China

Yancheng Institute of Technology, Yancheng, China

G. Tan (✉) · F. Feng · Y. Wang · D. Jiang
Hohai University, Nanjing, China
e-mail: gptan@hhu.edu.cn

H. Xun · Q. Wang
Yancheng Institute of Technology, Yancheng, China

B. Ye et al. (eds.), *2nd International Conference on 5G for Ubiquitous Connectivity*, EAI/Springer Innovations in Communication and Computing,
https://doi.org/10.1007/978-3-030-22316-8_7

access network are investigated without and with successive interference cancellation (SIC) [8]. At the same time, to address the mobile traffic offloading and resource allocation problem, decentralized traffic offloading scheme is designed to encourage each small-cell to achieves its own maximum utility [9]. Under the premise of high service quality, a new measurement for effective capacity is proposed to quantify the maximum sustainable data rate in heterogeneous ultra-dense distributed networks [10]. Based on the distributed chunk-based optimization algorithm, the power and subcarrier allocation problems are jointly optimized. The proposed algorithm can strike a balance between the complexity and performance in the multi-carrier Ultra-Dense Networks [11]. This literature reveals user location prediction-based cell discovery (ULPCD) scheme for user-centric ultra-dense network (UUDN). It mainly sends network deployment information to user equipment (UE) where terminal user enters cell coverage for the first time [12]. In Ref. [13], the authors further analyzed the limitation of MIMO and investigate the limitation with efficient interference management strategies. Under the Consideration of handover and interference for edge users, a user-centric transmission scheme is adopted, which uses ZF coding scheme to eliminate the interference [14]. Similarly, the user-centered inter-cellular interference Coordination is proposed. Users are required to request interference nulling for interference base stations in the range of interference coordination. And main interferences for each user are suppressed, which makes its performance better than existing base station clustering methods. However, the process of interference nulling is random, and the remaining interference may still have an adverse effect on user performance especially in the UDN scenarios [15]. Thus, the lowest possible complexity is explored to further improve user performance.

Therefore, a novel user-centric interference coordination (CPUCIC) scheme based on cluster and priority is presented. Its corresponding procedure is as follows. On the one hand, some interference is transformed as useful signal with CoMP technology. On the other hand, some other interference is set null with beamforming. As a result, main interference for objective user can be mitigated. And user performance is improved with CPUCIC scheme.

The remainder of this paper is organized as follows. Section 2 presents system model and problem formulation. In the Sect. 3, CPUCIC scheme is proposed to solve the problem. Simulation results are demonstrated for verification in Sect. 4. Finally, Sect. 5 concludes the paper and presents future work.

2 System Model and Problem Formulation

2.1 System Model

In this paper, UDN with inter-cell interference is shown in Fig. 1. Multiple-antenna base stations and corresponding single-antenna users are distributed randomly in a particular area. In detail, the number of base stations is denoted as LO_{BS} ($LO_{BS} = LO+1$, the 0th base station is the target base station, LO is the number of interfering

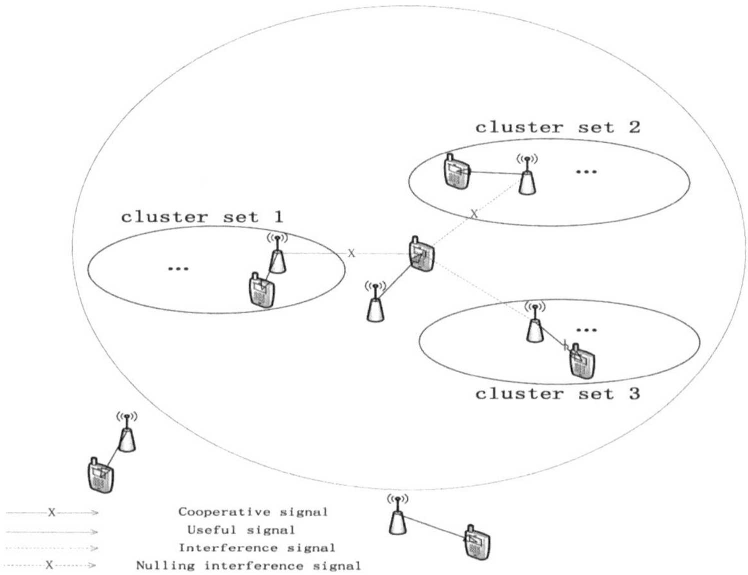

Fig. 1 System model

base stations). And the number of users is K. The service base station is assigned with the closest user, and the others are regarded as interfering base stations.All the interference BSs are divided into three clusters. The first cluster is the set of interference BSs with cooperation, which makes interference into useful signal. And the second cluster is the set of neighboring BSs with zero-forcing pre-coding technology. Then the third cluster is the remaining BSs. Each BS with multiple antennas forms a sub cluster. One of BSs is selected as cluster header to transmit data to objective user.

2.2 *Problem Formulation*

In the downlink of cellular communication network, the received signal for user k contains not only useful signal $SI_{k,i}$ from serving base station i, but also the interference signal $IN_{k,j}$ from adjacent base stations j ($j = 1, \ldots, LO_{BS}$ and $j \neq i$). The total signal for user k is denoted as $X_{k,i}$.

$$X_{k,i} = SI_{k,i} + \sum_{j=1, j \neq i}^{LO_{BS}} IN_{k,j} + N \tag{1}$$

where $SI_{k,i} = g_{k,i} h_{k,i} x_{k,i}$ is the desired signal for user k from base station i, $IN_{k,j} = g_{k,j} h_{k,j} x_{k,j}$ represents an interference signal for user k from the base station j, and N is white Gaussian noise. Besides, $g_{k,i} = p_i d_{k,i}^{-\alpha}$ represents large-scale fading of base station i to user k, p_i is transmission power from base station i, $d_{k,i}$ is the distance between base station i and user k, and α is attenuation factor. $h_{k,i}$ Represents the small-scale fading of base station i to user$k$, and $x_{k,i}$ represents the transmission signal of base station i.

In the model, the user ranks all the received interference power PO_i $(i = 1, \cdots, LO)$ from interference base stations in the descending order. It ranks results from the priority order of interference coordination, and $IN_1, IN_2 \ldots, IN_{LO}$ is obtained. After a series of user judgments, the minimum LO_1 and LO_2 can be determined by the conditions that meet $SINR \geq \gamma(\gamma$ is the threshold required for the minimum SINR of the user service). LO_1 represents the number of base stations that cooperate with serving base station BS_0 for the user, and LO_2 is denoted as the number of base stations BS_0 where serving base station is changed to the nulling beamforming.

It is assumed that the set of base station cooperating with serving base station and the set of base station with interference nulling determines the following parts Θ_1,Θ_2. At this time, signals in the set Θ_1 become to be useful for corresponding user. Meanwhile, signals in the set Θ_2 are still interference signals, which become to be zero by beamforming. The receiving signal of user is expressed as follows.

$$X_0 = SI_0 + \sum_{a \in \Theta_1} IN_{0,a} + \sum_{b \in \Theta_2} IN_{0,b} + \sum_{c \in \psi_{ICBS} - \Theta_1 - \Theta_2} IN_{0,c} + N \tag{2}$$

where $SI_0 + \sum_{a \in \Theta_1} IN_{0,a}$ is the useful signal received by the objective user, and two parts $\sum_{b \in \Theta_2} IN_{0,b} + \sum_{c \in \psi_{ICBS} - \Theta_1 - \Theta_2} IN_{0,c}$ is the interference for the objective user, the interference nulling $\sum_{b \in \Theta_2} IN_{0,b}$ is attained after serving base station beamforming, and $\sum_{c \in \psi_{ICBS} - \Theta_1 - \Theta_2} IN_{0,c}$ is the remaining interference. Therefore, the received signal of the user is expressed as follows.

$$X_0 = SI_0 + \sum_{a \in \Theta_1} IN_{0,a} + \sum_{c \in \psi_{ICBS} - \Theta_1 - \Theta_2} IN_{0,c} + N \tag{3}$$

At the point, the strong interference around the user UE_0 is converted into useful signal. A large part of the sub-strong interference is suppressed, which means that the remaining small part of interference no longer has any large impact on user performance. In addition, LO_1 and LO_2 which adjudicated by $SINR \geq \gamma$ are the smaller value during the user decision process. Therefore, the amount of information that the user UE_0 feeds back to local base station can also be reduced.

In this paper, we assume that all base stations have the same transmission power. Corresponding SINR for UE_0 is as follows.

$$SINR_k = \frac{SI_0 + \sum_{a \in \Theta_1} IN_{0,a}}{\sum_{c \in \psi ICBS - \Theta_1 - \Theta_2} IN_{0,c} + N} \tag{4}$$

$SI_0 + \sum_{a \in \Theta_1} IN_{0,a}$ is the power from its serving base station power and the other cooperative BSs. $\sum_{c \in \psi ICBS - \Theta_1 - \Theta_2} IN_{0,c} + N$ is the sum of interference base station power and noise.

The average SINR for users is as follows.

$$SINR_{avg} = \frac{1}{K} \sum_{k=1}^{K} SINR_k \tag{5}$$

The data rate of user is following.

$$R_k = B_k \log_2(1 + SINR_k) \tag{6}$$

Among them, the bandwidth B_k is allocated for the user k. Afterwards, we can obtain the average data rata of the user with following part.

$$R_{avg} = \frac{1}{K} \sum_{k=1}^{K} R_k \tag{7}$$

The total throughput in the system is following.

$$C_{sy} = \sum_{k=1}^{K} R_k \tag{8}$$

The spectral efficiency in the system is following.

$$\eta_{SE} = \frac{C_{sy}}{B_{sy}} \tag{9}$$

B_{sy} is denoted as the system bandwidth. It means the bandwidth is allocated by base station i for user k.

$$B_{k,i} = \frac{B_i}{K_i} \tag{10}$$

In this, K_i is the total number of users for BS_i, B_i is the bandwidth of the base station i, all base station share the same bandwidth.

In additions, a new performance parameter named average bandwidth gain-to-loss. Its expression is as follows.

$$\eta_{B_{k,i}} = \frac{B_{k,i}}{B_{k,i}^{non}} \cdot \left(\frac{SINR_{k,i}}{SINR_{k,i}^{non}} \right)^{-1} \tag{11}$$

where $B_{k,i}^{non}$ is the bandwidth of users and $SINR_{k,i}^{non}$ is the SINR under the NONCO scheme.

When $\frac{B_{k,i}}{B_{k,i}^{non}} > \frac{SINR_{k,i}}{SINR_{k,i}^{non}}$ the cost of the bandwidth for CPUCIC is at loss. When $\frac{B_{k,i}}{B_{k,i}^{non}} = \frac{SINR_{k,i}}{SINR_{k,i}^{non}}$, the cost of the bandwidth for CPUCIC is equal to gain. In other words, the cost of the bandwidth between loss and gain for CPUCIC reaches balance. When $\eta_{B_{k,i}}$ is less than 1, bandwidth losses are less than performance gains

$$\eta_B = \frac{\sum_K \eta_{B_{k,i}}}{K} \tag{12}$$

And $\sum_K \eta_{B_{k,i}}$ is the sum of the above performance for all users. η_B is average value in the whole system.

3 CPUCIC Scheme

The CPUCIC scheme mainly consists of three modules, including generation for coordination priorities, user decision procedures and cooperative transmission interference nulling process [16, 17]. The detailed description is as follows.

3.1 Generation for Coordination Priorities

The user firstly obtains the priority order of the interference coordination, according to the received signal power P. It is obtained from the surrounding LO interfering base stations $\Phi = \{IN_1, IN_2, \ldots, IN_{LO}\}$. Detailed description is shown in Fig. 2.

3.2 User Judgment Process

According to the coordination priority, the corresponding SINR is traversed successively. And the corresponding SINR is estimated, thus the minimum LO_1 and LO_2 that satisfy the conditions of $SINR \geq \gamma$ are calculated. LO_1 is the number of the BSs for cooperation in Θ_1. And LO_2 is the number of the BSs for interference nulling in Θ_2.

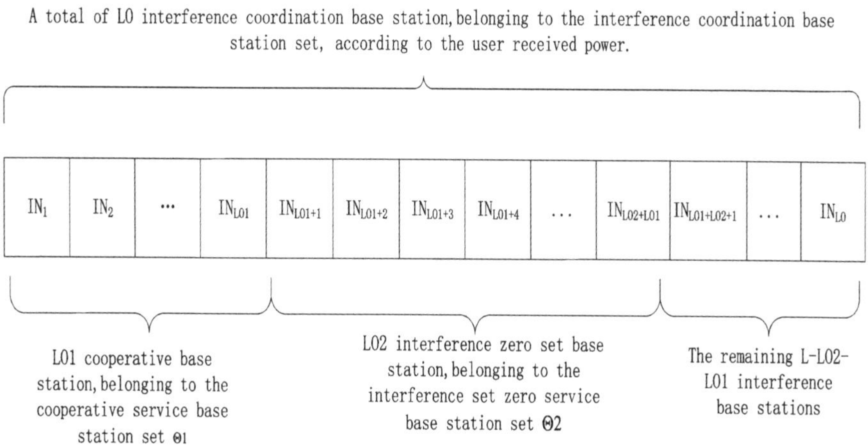

Fig. 2 Interference coordination priority

3.3 Collaborative Transmission and Interference Nulling Process

The LO_1 cooperative base stations and the LO_2 interference nulling base stations respectively implement the corresponding cooperative transmission [18] and interference nulling [19] for the objective user. At this point, the signal from the base station in the set Θ_1 becomes a useful signal for the UE_0. Meanwhile, the signal from the base station in the set Θ_2 is still an interference signal for the UE_0. However, the later signal can be set zero.

3.4 Corresponding Algorithm

Algorithm 1 System procedure

1. Input *LO1*, *LO2*.
2. Obtain Coordinate priority for each received power in Sect. 3.1
3. User determines *LO1* and *LO2* in Sect. 3.2.
4. *LO1* is used for coordination transmission, and *LO2* is used for interference nulling in Sect. 3.3.
5. Output *LO1*,*LO2*

4 CPUCIC Performance Analysis

In order to effectively evaluate the performances of CPUCIC, UCIIC and NONCO schemes are used for performance comparison. The UCIIC scheme randomly copes with interference in the optimal user-centered interference range. NONCO is a traditional interference suppression mechanism without base station collaboration and interference zero-forcing. The detailed comparable simulation is shown as follows.

4.1 Performance Analysis of CPUCIC in the Case of Different Base Stations

The corresponding simulation parameters are listed in Table 1, and Figs. 3, 4, 5 show the comparison of the average user SINR and the average user SE with different base station numbers. In addition, corresponding analysis for average bandwidth gain-to-loss rates is provided as follows.

As shown in Fig. 3, average users' SINR with CPUCIC scheme is above the 12dB while the number of BSs is different. There has an increase compared to the UCIIC scheme in the range of 8~22dB. Besides, CPUCIC scheme has an obvious increase compared to the NONCO scheme in the range of 7~14dB. The reason is as follows. No matter how many serving BSs and interfering BSs stay around the user, CPUCIC scheme can ensure its performance with BSs collaboration, when the condition $SINR \geq \gamma$ is met. Even the UCIIC scheme proceeds interference zero-forcing by determining the user-centric coordination range. Although the NONCO scheme can mitigate little strong interference, the SINR improvement of UCIIC scheme is far less than that of CPUCIC scheme. On one side, less strong interference from surrounding BSs generates and the user SINR can be easily improved by base station collaboration and interference zero-forcing when the number of BSs is small. On the other side, there has strong interference around objective users when the

Table 1 System parameters with different BSs

Parameter	Values
Radius of simulation area (m)	50
BS bandwidth BSBW (MHz)	20
SINR threshold γ (dB)	10
Base station antenna number T	4
Number of users $UENUM$	15
Number of BSs (m)	5–20

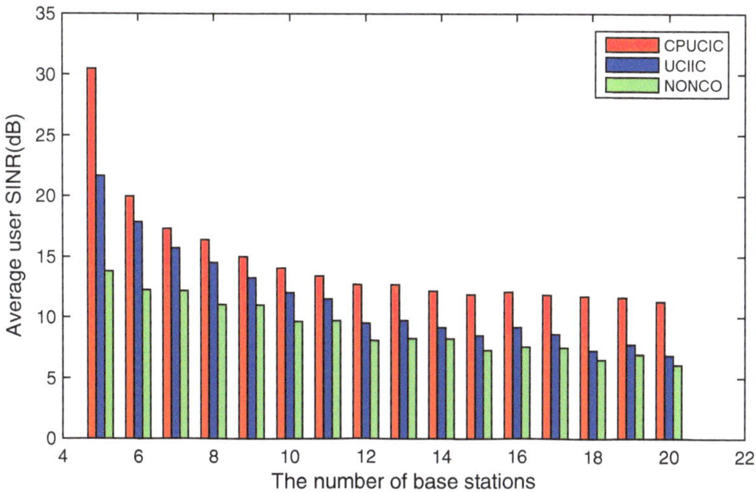

Fig. 3 The number of base stations vs. average user SINR (dB)

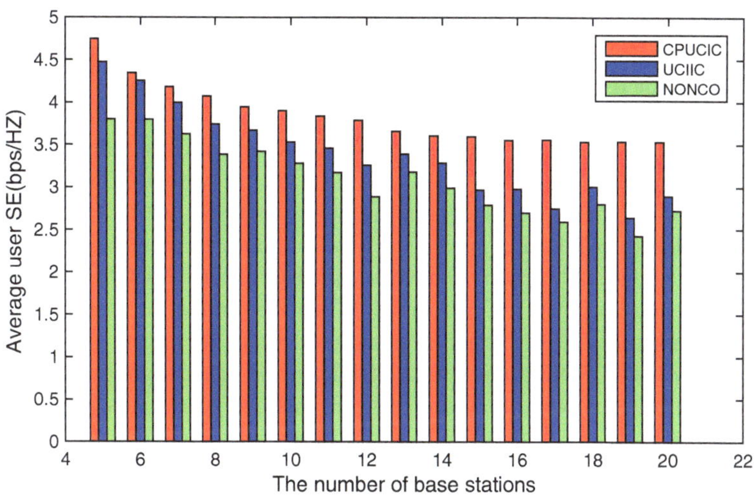

Fig. 4 The number of base stations vs. average user SE (bps/Hz)

number of BSs is large. If the user SINR will be kept at the same SINR level as mentioned above, you need more collaboration BSs to serve the user. Nevertheless, CPUCIC scheme will not take advantage of BS with no limitation. It will select the minimum L_1 to control complexity as low as possible, which satisfies the decision condition $SINR \geq \gamma$.

Consequently, the increasing number of BSs makes target user SINR reduced for the CPUCIC scheme. Nevertheless, it is not mainly actual cause. Considered from the fact, the increasing number of BSs produces higher complexity for

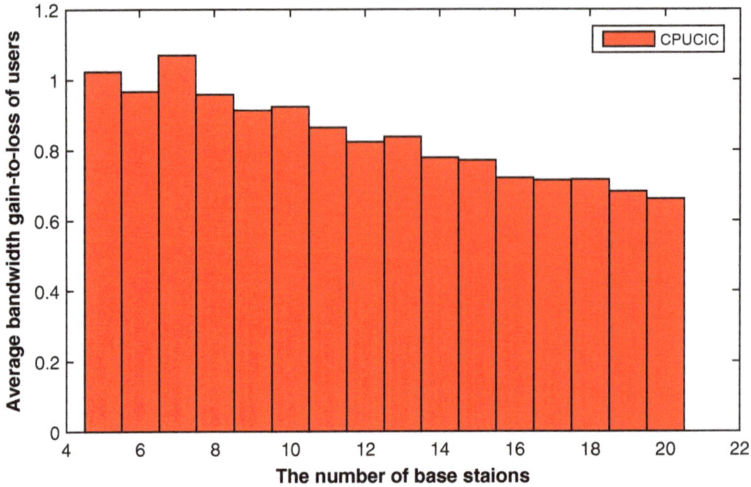

Fig. 5 The number of base stations vs. average bandwidth gain-to-loss of users

their collaboration, which leads SINR curve to tend to threshold. Thus, the BSs collaboration increase. It can be shown in Fig. 3 that the corresponding part has fallen.

As shown in Fig. 4, the average SE of NONCO scheme has a slight downward trend with the different BS numbers. The average SE with UCIIC scheme also has a slight downward trend. Nevertheless, the average SE with CPUCIC scheme decreased slightly while the number of BSs continues to increase. In the whole process, the SE basically remains above 3.5bps. UCIIC and CPUCIC schemes have obvious advantages for NONCO scheme. When the number of users is fixed, the increasing number of BSs may induce more interference for target users. According to Fig. 3, the users' SINR value for CPUCIC scheme has a slight downward trend and tends to be stable with the increasing BSs number. Afterwards, the bandwidth for each user is reduced, and the capacity of users tends to decrease. Due to two aforementioned analysis, the average SE with CPUCIC scheme tends to decrease and to be stable step by step. The user bandwidth remains unchanged in the other two mechanisms and the user SINR continues to decrease at the same time. Therefore, both of their user SE values has also been decreasing. Subsequently, the number of BSs continues to increase, and the SE of CPUCIC tends to be stable. It indicates that the advantages of CPUCIC at this stage begin to fully show. On account of the SINR differences between CPUCIC scheme and the other two schemes, the differences in SE become more and more obvious in the three schemes. User's SINR keeps above the SINR threshold, and the increase for total user capacity is larger than the increase of user bandwidth. Accordingly, CPUCIC scheme is equivalent to used limited bandwidth to exchange for larger SINR gain. And the user SE with CPUCIC scheme has been improved.

As shown in Fig. 5, the average bandwidth balance rate $\eta^B{}_{K,i}$ with CPUCIC scheme decreases when the number of BSs increases. The value of $\eta^B{}_{K,i}$ is less than 1 when the number of BSs is above 7. The value of $\eta^B{}_{K,i}$ is always greater than 1 when the number of BSs is between 5 and 7. The average bandwidth gain-to-loss $\eta^B{}_{K,i}$ <1, which means that the bandwidth loss is less than user performance gain. Furthermore, it is the gain for users. While the ratio $\eta^B{}_{K,i}$ >1 indicates that the bandwidth loss is greater than the user performance gain, that is the loss. In addition, the average bandwidth balance rate $\eta^B{}_{K,i}$ has a slight decrease and shows to be stable as the increasing BSs number continuously. It indicates that the average bandwidth gain-to-loss gradually has the tendency to draw near to turning point. The reason is as follows. More BSs there are, more serious interference exists around the objective user. To meet SINR threshold condition, the user requires more collaboration BSs. Therefore, the more bandwidth loss generates. As the number of BSs increases, SINR values is stably around the SINR threshold. The average bandwidth gain-to-loss $\eta^B{}_{K,i}$ will naturally decrease. This indicates that CPUCIC scheme attains the user SINR improvement at the cost of bandwidth loss. Consequently, the gain-to-loss is acceptable when the number of BSs is moderate, such as 8 to 20. As a result, CPUCIC scheme is effective and valuable.

Compared with the traditional interference suppression scheme UCIIC and NONCO, CPUCIC has improved effectively on user's SINR. As to the case of large number of BSs, although there are some limitations in terms of the number of antennas, CPUCIC effectively improves the user's SINR as well.

4.2 *Performance Analysis of CPUCIC in the Case of Changeable Users*

When the number of BSs is fixed, and more users appear in the wireless network. The corresponding parameters are set as shown in Table 2. Figures 6 and 7 show the comparison with the average SINR and user SE with the scheme CPUCIC, UCIIC,

Table 2 System parameters with different users

Parameter	Values
Radius of simulation area (m)	50
BS bandwidth BSBW (MHz)	20
SINR threshold γ (dB)	10
Base station antenna number T	4
Number of base stations $BSNUM$	40
Number of users $UENUM$ (m)	5–20

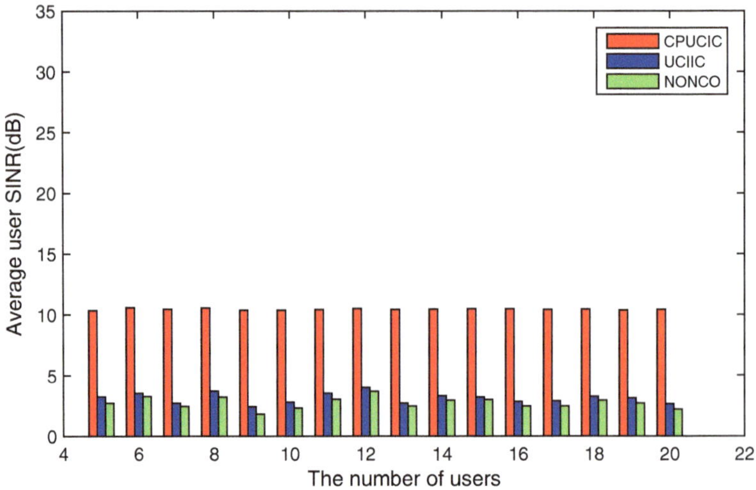

Fig. 6 The number of users vs. average user SINR (dB)

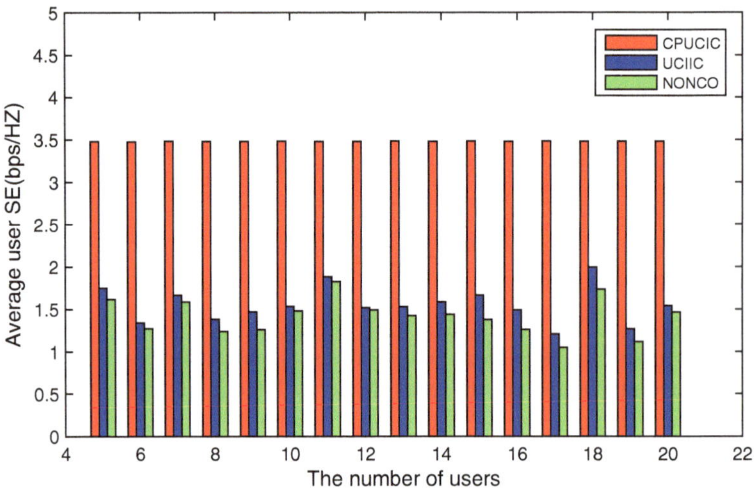

Fig. 7 The number of users vs. average user SE (bps/Hz)

and NONCO respectively in different numbers of users. In additions, the effect of base station number on the average bandwidth gain-to-loss of CPUCIC is analyzed.

First of all, the average SINR value for CPUCIC user approach to be 11dB stably in the case of different users in Fig. 6. The plots of UCIIC and NONCO fluctuate slightly in the range of 2.3∼4dB and 1.8∼3.8 dB respectively, but these two schemes do not vary greatly. The SINR threshold has been determined when the BSs number is not changed. The user SINR only fluctuate slightly around the SINR threshold in the scenario, which the number of BSs remains unchanged. Thus,

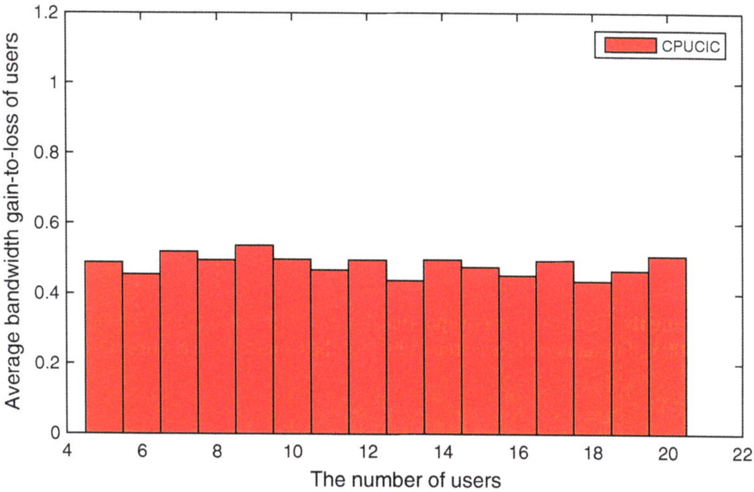

Fig. 8 The number of users vs. average bandwidth gain-to-loss ratio of users

the SINR performance remains stable. For the other two compared mechanisms, the number of BSs with strong interference is random. Consequently, the user SINR is also fluctuated randomly.

For UCIIC scheme, the allocated bandwidth for each user is decreased because of the increasing number of users. Nevertheless, the total bandwidth for all users remains invariable. As shown in Fig. 7, user SINR value with UCIIC scheme shows slight random fluctuation when the number of users increases. Corresponding user capacity shows slight random fluctuation too. The average SE fluctuation for UCIIC scheme is totally triggered by the fluctuation of its users' SINR. For NONCO scheme, the modification for the number of users has slight effect on the sum of users' SINR. Therefore, the average SE plot also shows a slight fluctuation.

Finally, the values of ${\eta^B}_{K,i}$ always have a slight fluctuation with more users for CPUCIC scheme in Fig. 8. The value of ${\eta^B}_{K,i}$ is always less than 0.6 while the number of users is in the range of 5∼20. $\eta_B > 1$ means that the bandwidth loss is greater than the user performance gain, which means loss. $\eta_B < 1$ shows that the user performance gain is greater than the bandwidth loss, which means gain. Some reasons are shown as follows. The number of users is small. And a small amount of bandwidth and SINR gain for each user can be obtained compared with NONCO scheme. Consequently, the average bandwidth gain-to-loss is at gain state at this stage. Afterwards, it leads to the result that the sum of the bandwidth has a slight fluctuation when the number of users increases. Consequently, the average balance bandwidth gain-to-loss ratio is at gain state. The user SINR with CPUCIC scheme is indeed improved, compared with the other two schemes. As a result, CPUCIC scheme improves the performance of users at the cost of bandwidth. On condition that the number of users is in the range of 5∼20, and such balance is acceptable. Furthermore, CPUCIC scheme is effectively valuable.

5 Conclusion

In this paper, the CPUCIC scheme is proposed for the UDN downlink interference, which significantly affects objective user performance. The CPUCIC scheme can further improve network performance, compared with UCIIC and NONCO scheme. However, the scheme is at the expense of more bandwidth consumption. The trade-off between bandwidth consumption and performance gain is deferred to future work.

Acknowledgements This work was supported by the National Natural Science Foundation of China (61832005), Fundamental Research Funds for the Central Universities (2014B33114).

References

1. Gao, M., Li, J., Jayakody, D.N.K., et al.: A super base station architecture for future ultra-dense cellular networks: toward low latency and high energy efficiency. IEEE Commun. Mag. **56**(6), 35–41 (2018)
2. Ge, X., Tu, S., Mao, G., et al.: 5G ultra-dense cellular networks. IEEE Wirel. Commun. **23**(1), 72–79 (2016)
3. Gotsis, A., Stefanatos, S., Alexiou, A.J.I.V.T.M.: UltraDense networks: the new wireless frontier for enabling 5G access. IEEE Veh. Technol. Mag. **11**(2), 71–78 (2016)
4. Kamel, M., Hamouda, W., Youssef, A.J.I.C.S., et al.: Ultra-dense networks: a survey. IEEE Commun. Surv. Tutorials **18**(4), 2522–2545 (2016)
5. Chang, Y., Yuan, X., Li, B., et al.: A joint unsupervised learning and genetic algorithm approach for topology control in energy-efficient ultra-dense wireless sensor networks. IEEE Commun. Lett. **22**(11), 2370–2373 (2018)
6. Wu, S., Wei, Y., Zhang, S., et al.: Gossip based distributed power control algorithm for 5G ultra dense networks. In: 2018 IEEE Wireless Communications and Networking Conference (WCNC), pp. 1–6 (2018)
7. Zhu, X., Xiao, C., Zeng, J., et al.: Virtual cell interference alignment in ultra dense network. In: 2016 IEEE/CIC International Conference on Communications in China (ICCC), pp. 1–6 (2016)
8. Shipeng, W., Huarui, Y., Guo, W.: Performance analysis of ultra dense network with linear reception and successive interference cancellation under limited backhaul. In: 2016 IEEE Conference on Computer Communications Workshops (INFOCOM WKSHPS), pp. 283–287 (2016)
9. Du, J., Gelenbe, E., Jiang, C., et al.: Contract design for traffic offloading and resource allocation in heterogeneous ultra-dense networks. IEEE J. Sel. Areas Commun. **35**(11), 2457–2467 (2017)
10. Cui, Q., Gu, Y., Ni, W., et al.: Preserving reliability of heterogeneous ultra-dense distributed networks in unlicensed spectrum. IEEE Commun. Mag. **56**(6), 72–78 (2018)
11. Shaozhen, G., Chengwen, X., Zesong, F., et al.: Distributed chunk-based optimization for multi-carrier ultra-dense networks. China Commun. **13**(1), 80–90 (2016)
12. Wu, H., Tian, H., Huang, Z., et al.: User location prediction based cell discovery scheme for user-centric ultra-dense networks. In: Wireless Communications and Networking Conference (WCNC), 2018 IEEE, pp. 1–6. IEEE, Piscataway (2018)
13. Liu, J., Sheng, M., Li, J.: Improving network capacity scaling law in ultra-dense small cell networks. IEEE Trans. Wirel. Commun. **17**(9), 6218–6230 (2018)

14. Song, S., Li, H., Fan, Y., et al.: Downlink interference rejection in ultra dense network. In: 2018 10th International Conference on Communication Software and Networks (ICCSN), pp. 361–364 (2018)
15. Li, C., Zhang, J., Haenggi, M., et al.: User-centric intercell interference nulling for downlink small cell networks. Trans. Wirel. Commun. **63**(4), 1419–1431 (2015)
16. Feng, F.: Research on Scheme of Interference Suppression for Downlink Ultra Dense Networks. Hohai Univisity (2016)
17. Tan, G., Wu, G., Li, Y., et al.: A downlink interference suppression scheme for ultra dense networks. In: International Conference on Computer Engineering and Networks, p. 023 (2017)
18. Cao, L., Hu, X., Zhang, M., et al.: Interactive CoMP with user-centric clustering based on load balancing in 5G dense networks. In: 2018 IEEE International Conference on Communications Workshops (ICC Workshops), pp. 1–6. IEEE, Piscataway (2018)
19. Gao, W., He, S., Chuai, G.: Clustering for coordinated zero-forcing beamformingin multi-user interference networks. In: 2016 IEEE International Conference on Network Infrastructure and Digital Content (IC-NIDC), pp. 322–326. IEEE, Piscataway (2016)

SAR Target Recognition via Enhanced Kernel Sparse Representation of Monogenic Signal

Chen Ning, Wenbo Liu, Gong Zhang, and Xin Wang

1 Introduction

As a significant means of remote sensing, synthetic aperture radar (SAR) has an important ability to produce high resolution images of surveillance, even in night and inclement weather conditions, due to its penetration capability and all-weather adaptability [1–3]. SAR automatic target recognition (ATR) has become a rapidly developing research field in radar post-processing and computer vision, and been studied thoroughly in the past several decades [4–6].

At present, the SAR ATR methods are mainly composed of three categories: template matching, model-based method and feature-based method. Template matching aims to search through the template library, which is generated by the training samples, to find the most matched template so as to obtain the classification label [7–10]. To handle these limitations, model-based methods for SAR ATR [11, 12] are addressed. Moreover, the model-based methods also include statistical type

C. Ning
College of Automation Engineering, Nanjing University of Aeronautics and Astronautics, Nanjing, China

School of Physics and Technology, Nanjing Normal University, Nanjing, China

W. Liu (✉)
College of Automation Engineering, Nanjing University of Aeronautics and Astronautics, Nanjing, China
e-mail: 48989221@qq.com

G. Zhang
Key Laboratory of Radar Imaging and Microwave Photonics, Ministry of Education, Nanjing University of Aeronautics and Astronautics, Nanjing, China

X. Wang
College of Computer and Information, Hohai University, Nanjing, China

B. Ye et al. (eds.), *2nd International Conference on 5G for Ubiquitous Connectivity*, EAI/Springer Innovations in Communication and Computing,
https://doi.org/10.1007/978-3-030-22316-8_8

[11] and physical type [12]. Recent studies on SAR ATR address feature-based method [13–17], which mainly has feature extraction step and classification step [18–20].

SAR ATR is still confronted with some challenges at present. One difficulty is that it is short of a compact, broad, and discriminative feature representation method for SAR targets [21]. Another challenge is how to design a robust and reliable classification algorithm when the features of different categories are not linearly separable. In the last few years, Felsberg [22] has proposed a new feature extraction method named monogenic signal, which combines the original 2-D signal with its Riesz transformation result to produce a new complicated analytic signal [23]. After introducing the multi-scale bandpass filter, the monogenic signal could be expressed by three orthogonal components, namely local phase, amplitude and orientation. Such decomposition lets monogenic signal be able to capture both the spatial information and spectral feature of SAR images at the same time [24]. Furthermore, to design a reliable classifier for the general target recognition problems, [25, 26] have present a kernel sparse representation-based classification method (KSRC). It utilizes kernel trick to map the original data which are non-linear into another feature space, which is sometimes higher dimensional. Thus, in the new space, the data of different classes can be linearly separable.

We present a novel SAR ATR algorithm based on the enhanced kernel sparse representation of monogenic signal here. Our main contributions include two aspects: (1) To capture the spectral and spatial features of a target at the same time, a multi-scale monogenic feature extraction scheme is proposed for SAR targets. (2) An enhanced kernel sparse representation-based classification method (KSRC) is designed. Different from the traditional KSRC, in the enhanced KSRC, we first integrate the KPCA as well as the kernel fisher discriminant analysis (KFDA) to generate an augmented pseudo-transformation matrix. Then, a new discriminative feature mapping approach is presented by exploiting the augmented pseudo-transformation matrix so that the feature dimension of the kernel feature space can be effectively reduced. At last, the ℓ_1-norm minimization is utilized to calculate sparse coefficients for a test sample, and the inference can be obtained by total reconstruction error minimization.

2 Related Work

2.1 The Monogenic Feature

Suppose $f(\mathbf{z})$ be a 2-D image and $f_R(\mathbf{z})$ be its Riesz-transformation result. Here, $\mathbf{z} = (x, y)^{\mathrm{T}}$ is the 2-D coordinate vector of spatial domain. Then, we can get the monogenic signal $f_M(\mathbf{z})$:

$$f_M(\mathbf{z}) = f(\mathbf{z}) - (i, j) f_R(\mathbf{z}) \tag{1}$$

Here, i and j are the imagery units [24]. After decomposition, we can compute the three orthogonal components of monogenic feature for the original 2-D images, namely, local amplitude, phase, and orientation respectively.

$$\begin{cases} amplitude: & A(\mathbf{z}) = \sqrt{f(\mathbf{z})^2 + |f_R(\mathbf{z})|^2} \\ phase: & \varphi(\mathbf{z}) = atan\left(|f_R(\mathbf{z})|, f(\mathbf{z})\right) \in (-\pi, \pi] \\ orientation: & \theta(\mathbf{z}) = atan\left(f_y(\mathbf{z}) / f_x(\mathbf{z})\right) \in \left(-\frac{\pi}{2}, \frac{\pi}{2}\right] \end{cases} \tag{2}$$

In (2), the local amplitude, phase and orientation represent the local energetic, orientation and structural features for the original images, respectively.

2.2 Kernel Sparse Representation

Suppose $\{\mathbf{x}_i, y_i\}_{i=1}^n$ be the training set with n training samples. Here, $\mathbf{x}_i \in \chi \subset \mathrm{R}^m$ and $y_i \in \{1, 2, \ldots, c\}$ with c-class totally. If there exists a nonlinear map function Φ, which is implicitly linked to a kernel function $k(\cdot, \cdot)$, we can utilize Φ to map the data of the original nonlinear space to a kernel feature space $\mathcal{F}$ [26]. Thus, the training samples can be linearly separable in the new space.

$$\Phi : \mathbf{x} \in \chi \rightarrow \Phi(\mathbf{x}) = [\phi_1(\mathbf{x}), \phi_2(\mathbf{x}), \ldots, \phi_D(\mathbf{x})]^T \in \mathcal{F} \tag{3}$$

Here, $\Phi(\mathbf{x}) \in \mathrm{R}^D$ indicates the image of $\mathbf{x}$, with $D \gg m$ being the dimension of the new kernel space. So, the images of the training samples $\mathbf{x}_i$ are $\Phi(\mathbf{x}_i)$, $i = 1, \ldots, n$. Then, according to the sparse representation theory, the image of test sample can be represented by linearly combining the ones of training set in the kernel space:

$$\Phi(\mathbf{x}) = \sum_{i=1}^{n} \alpha_i \Phi(\mathbf{x}_i) = \mathbf{\Phi}\boldsymbol{\alpha} \tag{4}$$

Here, $\boldsymbol{\alpha} = [\alpha_1, \alpha_2, \ldots, \alpha_n]^T$ is the sparse representation coefficient vector, which holds the coefficients for the images $\Phi(\mathbf{x}_i)$. So, we can get the new image matrix:

$$\mathbf{\Phi} = [\Phi(\mathbf{x}_1), \Phi(\mathbf{x}_2), \ldots, \Phi(\mathbf{x}_n)] \in \mathrm{R}^{D \times n} \tag{5}$$

We can get the new sparse representation optimization problem by replacing the SRC constraint by (4)

$$\min_{\boldsymbol{\alpha}} \|\boldsymbol{\alpha}\|_1 \quad \text{subject to} \quad \Phi(\mathbf{x}) = \mathbf{\Phi}\boldsymbol{\alpha} \tag{6}$$

However, Φ is often unknown and it is difficult to solve the (6) optimization problem directly. Yet, it is fortunate for us to be able to utilize the kernel trick and kernel-based dimension reduction methods to turn (6) to a feasible optimization problem. Suppose $\mathbf{P} \in \mathrm{R}^{D \times d}$ be the transformation matrix. By multiplying it to both sides of (4), we can get the formula:

$$\mathbf{P}^T \Phi(\mathbf{x}) = \mathbf{P}^T \boldsymbol{\Phi}\boldsymbol{\alpha} \tag{7}$$

Inspired by the idea of kernelized dimension reduction methods such as KPCA [27] and KFDA [28], we can use the linear combination of training set images to represent the column vector $\mathbf{P}_j$ of transformation matrix $\mathbf{P} = [\mathbf{P}_1, \ldots, \mathbf{P}_d]$, i.e., the projection vector:

$$\mathbf{P}_j = \sum_{i=1}^{n} \beta_{j,i} \Phi(\mathbf{x}_i) = \boldsymbol{\Phi}\boldsymbol{\beta}_j \tag{8}$$

Here, $\boldsymbol{\beta}_j = [\beta_{j,1}, \ldots, \beta_{j,n}]^T$ and $\mathbf{B} = [\boldsymbol{\beta}_1, \ldots, \boldsymbol{\beta}_d]$ are the pseudo-transformation vector and pseudo-transformation matrix respectively. So, we can get the transformation matrix:

$$\mathbf{P} = \boldsymbol{\Phi}\mathbf{B} \tag{9}$$

Substituting (9) into (7), we get

$$\mathbf{B}^T \mathbf{k}(\cdot, \mathbf{x}) = \mathbf{B}^T \mathbf{K}\boldsymbol{\alpha} \tag{10}$$

where $\mathbf{k}(\cdot, \mathbf{x}) = [k(\mathbf{x}_1, \mathbf{x}), \ldots, k(\mathbf{x}_n, \mathbf{x})]^T = \boldsymbol{\Phi}^T \Phi(\mathbf{x})$, $\mathbf{K} = \boldsymbol{\Phi}^T \boldsymbol{\Phi} \in \mathrm{R}^{n \times n}$ is the kernel matrix and its elements $K_{ij} = k(\mathbf{x}_i, \mathbf{x}_j)$. In order to finding the pseudo-transformation matrix $\mathbf{B}$, [26] describes the schemes such as KPCA and KFDA.

3 Presented Method

3.1 Multi-Scale Monogenic Feature Extraction

Since the monogenic feature has the good feature of capturing both spatial and spectral information at the same time [29], we adopt multi-scale monogenic signal analysis method to extract the feature of the SAR target.

First, by employing the bandpass Log-Gabor filter [30], we can rewrite the formula of monogenic signal (1) as:

$$f_M(\mathbf{z}) = \left(h_{lg}(\mathbf{z}) * f(\mathbf{z})\right) - (i, j)\left(h_{lg}(\mathbf{z}) * f_R(\mathbf{z})\right) \tag{11}$$

Here, h_{lg} is the Log-Gabor function. Moreover, through tuning the scale of Log-Gabor bandpass filter [32], we obtain the multi-scale monogenic signal

$\{f_M^1, \cdots, f_M^S\}$, along with the number of scales S and the monogenic signal of the kth scale f_M^k. Thus, we can get the multi-scale monogenic signal components:

$$\left\{\underbrace{A^1, \phi^1, \theta^1}_{f_M^1}, \cdots, \underbrace{A^S, \phi^S, \theta^S}_{f_M^S}\right\} \tag{12}$$

where A^k, ϕ^k, and θ^k ($k = 1, \ldots, S$) indicate the local amplitude, phase, and orientation of the kth scale monogenic signal.

Second, in order to make the multi-scale monogenic feature realistic to be applied to the subsequent processing, this paper reshapes and concatenates the monogenic components to form a new integral feature vector,

$$\begin{array}{l} \chi_A^k = vec\left(A^k\right) \quad \chi_\phi^k = vec\left(\phi^k\right) \quad \chi_\theta^k = vec\left(\theta^k\right) \\ \chi = \left[\chi_A^1; \cdots; \chi_A^S; \chi_\phi^1; \cdots; \chi_\phi^S; \chi_\theta^1; \cdots; \chi_\theta^S\right] \end{array} \tag{13}$$

Here, the symbol $vec(\cdot)$ demonstrates the operation of converting matrix to vector. χ is the resulting monogenic feature vector.

3.2 Pseudo-Transformation Matrix and Dictionary Construction

In order to make the feature more discriminative and further improve the classification accuracy, in this section, we propose to combine KPCA and KFDA to construct an augmented pseudo-transformation matrix.

First, we compute the vectors of pseudo-transformation $\boldsymbol{\beta}_j$ via KPCA scheme, which are the normalized eigenvectors solved from the eigenvalue problem:

$$\mathbf{K}\boldsymbol{\beta} = \lambda\boldsymbol{\beta} \tag{14}$$

By computing the d largest eigenvalues $\lambda_j, j = 1, \ldots, d$, where $\lambda_1 \geq \lambda_2 \geq \cdots \geq \lambda_d$, and taking their corresponding eigenvectors, we can obtain the KPCA pseudo-transformation matrix $\mathbf{B}' = [\boldsymbol{\beta}_1, \ldots, \boldsymbol{\beta}_d] \in \mathrm{R}^{n \times d}$.

Second, we obtain another pseudo-transformation matrix $\mathbf{B}'' \in \mathrm{R}^{n \times d}$ by KFDA scheme, which is the solution for this maximization problem:

$$\max_{\mathbf{B}} \frac{tr\left(\mathbf{B}^T \mathbf{S}_b^K \mathbf{B}\right)}{tr\left(\mathbf{B}^T \mathbf{S}_w^K \mathbf{B}\right)} \tag{15}$$

where $tr(\cdot)$ demonstrates the matrix trace. $\mathbf{S}_w^K$ and $\mathbf{S}_b^K$ denotes the within-class and between-class scatter matrices, respectively. In general, we have $d < c$.

Thirdly, in order to combine both the advantages of KPCA and KFDA, we build an augmented pseudo-transformation matrix as $\mathbf{B} = [\mathbf{B}^{'}, \mathbf{B}'']$. Consequently, the final dictionary in kernel feature space $\mathcal{F}$ is constructed by $\mathbf{B}^T\mathbf{K}$.

3.3 KSRC Optimization Problem Solver and Classifier

In the query stage, we transform the query SAR image and extract its monogenic features of three scales as (13). Then we feed them into the KSRC optimization problem solver module, where the $\mathbf{k}(\cdot, \mathbf{x})$ vector is computed at first.

Subsequently, the sparse representation coefficients in the kernel feature space will be computed via the KSRC optimization problem solver, with such parameters as the pseudo-transformation matrix $\mathbf{B}$ and dictionary in kernel feature space $\mathbf{B}^T\mathbf{K}$ transferred from the dictionary construction stage. If the constraint in (6) is replaced by (10), a new optimization problem can be obtained:

$$\min_{\alpha} \|\alpha\|_1 \quad \text{subject to} \quad \mathbf{B}^T\mathbf{k}(\cdot, \mathbf{x}) = \mathbf{B}^T\mathbf{K}\alpha \tag{16}$$

By solving (16), the sparse representation vector $\boldsymbol{\alpha}$ is computed. Then, we can utilize $\boldsymbol{\alpha}$ to classify $\mathbf{x}$.

Through constructing a new vector $\boldsymbol{\delta}_i$ that only hold the coefficients corresponding to class i, we get the approximation of the ith class for the query sample $\mathbf{x}$, which is formed as $\mathbf{B}^T\mathbf{K}\boldsymbol{\delta}_i$. Eventually, we compute the residual between $\mathbf{B}^T\mathbf{k}(\cdot, \mathbf{x})$ and its approximations, and predict the label $\hat{y}$ for $\mathbf{x}$ via the rule of total reconstruction error minimization as follows

$$\hat{y} = \arg \min_{i=1,\ldots,c} r_i(\mathbf{x}) = \left\|\mathbf{B}^T\mathbf{k}(\cdot, \mathbf{x}) - \mathbf{B}^T\mathbf{K}\boldsymbol{\delta}_i\right\|_2 \tag{17}$$

4 Experimental Results

4.1 Experimental Setup

Our proposed method is evaluated by a great number of experiments based on the public moving and stationary target acquisition and recognition dataset (MSTAR) [31]. This gallery is collected by a spotlight mode SAR sensor, which is operated in one-foot resolution and X-band. We provide the data details of our experiments in Table 1. Specifically, we totally use four target classes for test, which are

Table 1 Summary of the training and testing images used for experiments

Depression angle	BMP2	T72	BTR60	T62	Total
17° [Training]	233 (SN_9563)	232 (SN_132)	256	299	1020
15° [Testing]	196 (SN_9566)	195 (SN_812)	195	273	1246
	196 (SN_C21)	191 (SN_S7)			

Table 2 Algorithms to be compared with ours

Algorithm	Description
KSRC-M	Enhanced kernel sparse representation classification method with monogenic features (Our combined KPCA and KFDA scheme)
KSRC-M^1	Kernel sparse representation classification method with monogenic features (Only KPCA scheme)
KSRC-M^2	Kernel sparse representation classification method with monogenic features (Only KFDA scheme)
TJSRC-M	Tritask joint sparse representation classification method with monogenic features
SRC-M	Sparse representation classification method with monogenic features
SVM-M	Linear support vector machine classification method with monogenic features
KSRC-I	Kernel sparse representation classification method (Combined KPCA and KFDA scheme) with intensity feature
SRC-I	Sparse representation classification method with intensity feature

BMP2, T72, BTR60, and T62 respectively. Thereinto, T72 and BMP2 have some manufacture versions, which have several modifications in structure (expressed via serial numbers). In the table, we show the number of aspect view for different target, and indicate their series number in the bracket. For each target class, images were captured at two different depression angles (i.e., 17° and 15°). In our experiments, we utilize the images of target collected at 17° depression angle for training set, and those captured at 15° depression angle for testing set. Besides, all the target images are cropped to 64×64 pixels.

We test our proposed method with seven benchmark algorithms for the SAR ATR accuracy. Table 2 summarizes the eight algorithms. Thereinto, the KSRC-M is our presented method, which contains our combined KPCA and KFDA scheme. KSRC-M^1 is the kernel sparse representation classifier with monogenic features, which only contains the KPCA scheme. KSRC-M^2 is the kernel sparse representation classifier with monogenic features, which only contains the KFDA scheme. In addition, TJSRC-M represents the method in [32], which introduced a tri-task joint sparse representation method with monogenic signal for SAR target recognition. SRC-M represents the method that utilizes the sparse representation classification method with monogenic features [21]. SVM-M employees the linear support vector machine to make the inferences [18]. KSRC-I represents the method that uses our kernel sparse representation classifier with intensity features. SRC-I is the sparse representation classification method with intensity feature. All methods are implemented in the same experimental environment as well as training/test sets.

4.2 Target Recognition Results

Table 3 shows the experimental results about confusion matrices and overall recognition rates. We can see that our proposed KSRC-M method achieves the best performance for SAR target recognition. Compared with KSRC-M^1 and KSRC-M^2, KSRC-M has higher overall recognition rates. These results are not surprising, for KSRC-M is an augmented recognition framework that combines the advantages of the monogenic feature, KPCA and KFDA. Furthermore, compared with other state-of-the-art algorithms, our KSRC-M also has better results, as it can precisely depict the nonlinear characteristics of data, and accurately predict the class label of the query.

Table 3 The results of four target recognition

	KSRC-M (0.9462)				KSRC-M^1 (0.9374)			
	BMP2	T72	BTR60	T62	BMP2	T72	BTR60	T62
BMP2	0.9541	0.0306	0.0077	0.0077	0.9668	0.0230	0.0077	0.0026
T72	0.0104	0.8808	0	0.1088	0.0130	0.8394	0	0.1477
BTR60	0	0	0.9897	0.0103	0	0	0.9949	0.0051
T62	0	0.0037	0	0.9963	0	0.0073	0	0.9927
	KSRC-M^2 (0.8339)				TJSRC-M (0.9117)			
	BMP2	T72	BTR60	T62	BMP2	T72	BTR60	T62
BMP2	0.8393	0.0536	0.0281	0.0791	0.9107	0.0510	0.0230	0.0153
T72	0.0699	0.7358	0.1503	0.0440	0.0181	0.8394	0	0.1425
BTR60	0.1128	0.0154	0.8564	0.0154	0.0256	0.0103	0.9385	0.0256
T62	0.0476	0	0.0037	0.9487	0	0.0037	0	0.9963
	SRC-M (0.9053)				SVM-M (0.8868)			
	BMP2	T72	BTR60	T62	BMP2	T72	BTR60	T62
BMP2	0.9311	0.0536	0.0051	0.0102	0.8546	0.1071	0.0255	0.0128
T72	0.0363	0.7720	0.0078	0.1839	0.0829	0.8342	0.0052	0.0777
BTR60	0	0.0051	0.9949	0	0.0308	0.0051	0.9231	0.0410
T62	0	0.0073	0	0.9927	0	0.0183	0	0.9817
	KSRC-I (0.8900)				SRC-I (0.8708)			
	BMP2	T72	BTR60	T62	BMP2	T72	BTR60	T62
BMP2	0.8036	0.1301	0.0357	0.0306	0.9617	0.0255	0.0102	0.0026
T72	0.0233	0.8523	0.0026	0.1218	0.0466	0.6477	0.0078	0.2979
BTR60	0	0	1	0	0.0154	0	0.9692	0.0154
T62	0	0.0073	0.0037	0.9890	0	0.0147	0	0.9853

5 Conclusion

The main contribution of our paper is the design of a unified framework that integrates the multi-scale monogenic signal and an enhanced KSRC classifier. In the enhance KSRC, an augmented pseudo-transformation matrix is generated by combining KPCA and KFDA. By using such augmented pseudo-transformation matrix, a more discriminative feature mapping can be achieved. Based on it, the dimension of the kernel feature space can be effectively reduced. Finally, ℓ_1-norm minimization is utilized to calculate the sparse coefficients for a test sample, and thus the inference can be reached by the rule of minimizing total reconstruction error. Compared with many state-of-the-art SAR ATR algorithms, our proposed method has the best performance. Future work includes testing our method under more different operation conditions.

Acknowledgement This work was supported by the National Natural Science Foundation of China (Grant No. 61603124, 61871218, 61801211, 61501233), Funding of Jiangsu Innovation Program for Graduate Education (Grant No. KYLX15_0278), Fundamental Research Funds for the Central Universities (Grant No. 2019B15314, 3082017NP2017421), and the Aeronautical Science Foundation of China (Grant No. 20152052026).

References

1. Keydel, E.R., Lee, S.W., Moore, J.T.: MSTAR extended operating conditions: a tutorial. Proc. SPIE. **2757**, 228–242 (1996)
2. López-Martínez, C., Pottier, E.: Coherence estimation in synthetic aperture radar data based on speckle noise modeling. Appl. Opt. **46**(4), 544–558 (2007)
3. Qi, F., Ocket, I., Schreurs, D., Nauwelaers, B.: A system-level simulator for indoor mmW SAR imaging and its applications. Opt. Express. **20**(21), 23811–23820 (2012)
4. Zhang, H., Nasrabadi, N.M., Zhang, Y., Huang, T.S.: Multi-view automatic target recognition using joint sparse representation. IEEE Trans. Aerosp. Electron. Syst. **48**(3), 2481–2497 (2012)
5. Ross, T.D., Worrell, S.W., Velten, V.J., Mossing, J.C., Bryant, M.L.: Standard SAR ATR evaluation experiments using the MSTAR public release data set. Proc. SPIE. **3370**, 566–573 (1998)
6. Huan, R., Pan, Y.: Decision fusion strategies for SAR image target recognition. IET Radar Sonar Navig. **5**(7), 747–755 (2011)
7. Owirka, G.J., Verbout, S.M., Novak, L.M.: Template-based SAR ATR performance using different image enhancement techniques. Proc. SPIE. **3721**, 302–319 (1999)
8. Casasent, D., Nehemiah, A.: Confuser rejection performance of EMACH filters for MSTAR ATR. Proc. SPIE. **6245**, 62450D (2006)
9. Liu, M., Wu, Y., Zhang, Q., Wang, F., Li, M.: Synthetic aperture radar target configuration recognition using locality-preserving property and the Gamma distribution. IET Radar Sonar Navig. **10**(2), 256–263 (2016)
10. Park, J., Kim, K.: Modified polar mapping classifier for SAR automatic target recognition. IEEE Trans. Aerosp. Electron. Syst. **50**(2), 1092–1107 (2014)
11. DeVore, M.D., O'Sullivan, J.A.: Quantitative statistical assessment of conditional models for synthetic aperture radar. IEEE Trans. Image Process. **13**(2), 113–125 (2004)

12. Zhou, J., Shi, Z., Cheng, X., Fu, Q.: Automatic target recognition of SAR images based on global scattering center model. IEEE Trans. Geosci. Remote Sens. **49**(10), 3713–3729 (2011)
13. Akbarizadeh, G.: A new statistical-based kurtosis wavelet energy feature for texture recognition of SAR images. IEEE Trans. Geosci. Remote Sens. **50**(11), 4358–4368 (2012)
14. Liu, X., Huang, Y., Pei, J., Yang, J.: Sample discriminant analysis for SAR ATR. IEEE Geosci. Remote Sens. Lett. **11**(12), 2120–2124 (2014)
15. Cui, Z., Cao, Z., Yang, J., Feng, J., Ren, H.: Target recognition in synthetic aperture radar images via non-negative matrix factorisation. IET Radar Sonar Navig. **9**(9), 1376–1385 (2015)
16. Amoon, M., Rezai-rad, G.: Automatic target recognition of synthetic aperture radar (SAR) images based on optimal selection of Zernike moments features. IET Comput. Vis. **8**(2), 77–85 (2014)
17. Srinivas, U., Monga, V., Raj, R.G.: SAR automatic target recognition using discriminative graphical models. IEEE Trans. Aerosp. Electron. Syst. **50**(1), 591–606 (2014)
18. Zhao, Q., Principe, J.: Support vector machines for SAR automatic target recognition. IEEE Trans. Aerosp. Electron. Syst. **37**(2), 643–654 (2001)
19. Sun, Y., Liu, Z., Todorovic, S., Li, J.: Adaptive boosting for SAR automatic target recognition. IEEE Trans. Aerosp. Electron. Syst. **43**(1), 112–125 (2007)
20. Yang, S., Ma, Y., Wang, M.: Compressive feature and kernel sparse coding-based radar target recognition. IET Radar Sonar Navig. **7**(7), 755–763 (2013)
21. Dong, G., Wang, N., Kuang, G.: Sparse representation of monogenic signal: with application to target recognition in SAR images. IEEE Signal Process. Lett. **21**(8), 952–956 (2014)
22. Felsberg, M., Sommer, G.: The monogenic signal. IEEE Trans. Signal Process. **49**(12), 3136–3144 (2001)
23. Felsberg, M., Sommer, G.: The monogenic scale-space: a unifying approach to phase-based image processing in scale-space. J. Math. Imaging Vision. **21**(1-2), 5–26 (2004)
24. Dong, G., Kuang, G., Zhao, L., Lu, J., Lu, M.: Joint sparse representation of monogenic components: with application to automatic target recognition in SAR imagery. Proc. IEEE Symp. Geosci. Remote Sens. **8**(7), 549–552 (2014)
25. Gao, S., Tsang, I.W.-H., Chia, L.-T.: Sparse representation with kernels. IEEE Trans. Image Process. **22**(2), 423–434 (2013)
26. Zhang, L., Zhou, W.-D., Chang, P.-C., Liu, J., Yan, Z.: Kernel sparse representation-based classifier. IEEE Trans. Signal Process. **60**(4), 1684–1695 (2012)
27. Schölkopf, B., Smola, A.J., Müller, K.-R.: Nonlinear component analysis as a kernel eigenvalue problem. Neural Comput. **10**(5), 1299–1319 (1998)
28. Mika, S., Rätsch, G., Weston, J., Schölkopf, B., Müller, K.-R.: Fisher discriminant analysis with kernels. In: Proceedings of the 1999 IEEE Signal Processing Society Workshop, Neural Networks for Signal Processing IX (Cat. No.98TH8468), pp. 41–48 (1999)
29. Yang, M., Zhang, L., Shiu, S.C.-K., Zhang, D.: Monogenic binary coding: an efficient local feature extraction approach to face recognition. IEEE Trans. Inf. Forensics Secur. **7**(6), 1738–1751 (2012)
30. Dong, G., Kuang, G.: Classification on the monogenic scale space: application to target recognition in SAR image. IEEE Trans. Image Process. **24**(8), 2527–2539 (2015)
31. Mossing, J.C., Ross, T.D.: An evaluation of SAR ATR algorithm performance sensitivity to MSTAR extended operating conditions. Proc. SPIE. **3370**, 554–565 (1998)
32. Dong, G., Kuang, G., Wang, N., Zhao, L., Lu, J.: SAR target recognition via joint sparse representation of monogenic signal. IEEE J. Sel. Top. Appl. Earth Observ. Remote Sens. **8**(7), 3316–3328 (2015)

Optimization of FBMC Waveform by Designing NPR Prototype Filter with Improved Stopband Suppression

Jingyu Hua, Jiangang Wen, Anding Wang, Zhijiang Xu, and Feng Li

1 Introduction

Mobile communication will be further developed to support the advent of the Internet of Things, which requires a significant improvement in data rate, latency, energy cost and number of connections [1–3]. As determining the required technique in five generation communications, the waveform design has been greatly studied to fulfill these requirements. In the unified frame proposed by 5GNOW, several waveforms are combined to handle with different types of traffic [4, 5]. Among them, FBMC can reduce the high out-of-band emission in OFDM and is prominent for its high time-frequency efficiency [6]. Moreover, FBMC has the least sensitivity to synchronization errors in contrast to other multicarrier technics like UFMC and F-OFDM [7].

FBMC can be viewed as an evolution of OFDM, which applies high quality filters in each subcarrier. Benefiting from the flexible filter design, the sidelobes of FBMC can be greatly suppressed, making FBMC a good candidate for asynchronous multiple access and fragmented spectrum communications [8, 9]. Furthermore, to guarantee the perfect reconstruction (PR) of FBMC signal, the generalized Nyquist condition along both time and frequency axises should be satisfied by FBMC prototype filter [10]. However, this PR property can only be obtained in ideal channel, thus the NPR prototype filter is preferred in practical FBMC systems. When the underlying channel is time-invariant or varies slowly, some widely used

J. Hua (✉) · A. Wang
School of Information and Electronic Engineering, Zhejiang Gongshang University, Hangzhou, China

J. Wen · Z. Xu · F. Li
College of Information Engineering, Zhejiang University of Technology, Hangzhou, China

B. Ye et al. (eds.), *2nd International Conference on 5G for Ubiquitous Connectivity*, EAI/Springer Innovations in Communication and Computing,
https://doi.org/10.1007/978-3-030-22316-8_9

square-root Nyquist filters have been presented, otherwise well-localized prototype filters are preferred [11].

Early works of FBMC prototype filter had provided various kinds of orthogonal pulse in continuous-time [12, 13]. Among, the isotropic orthogonal transform algorithm (IOTA), which could be acquired by applying a two-step orthogonalization to the Gaussian function, was popular for its high time-frequency localization [14]. Another popular filter was the PHYDYAS filter, which was designed by frequency sampling method in discrete time and had been employed for its superior sidelobe falling rate [15]. However, the IOTA filter and the PHYDYAS filter could not effectively relax Nyquist condition for better sidelobe suppression. To tackle this issue, some direct optimization methods had been proposed. In [16, 17], the iterative least-square (LS) method and the iterative second-order cone programming (SOCP) method had been applied to realize the tradeoff between Nyquist condition and stopband energy, but the control was made by inefficient weightings.

Different from the methods discussed above, our proposed method can straightly control the Nyquist condition under the object of minimizing stopband energy. In order to circumvent the nonconvex problem in Nyquist condition, the original constraints are transformed to linear ones, in which the filter autocorrelation coefficients are taken as the new variables for optimization. Correspondingly, the cost function and other constraints are also transformed to linear expressions, in which the Nyquist condition can be easily relaxed for better sidelobe suppression. Once the autocorrelation coefficients are obtained, a spectral factorization [18] is employed to retrieve the final NPR prototype filter.

In the proposed method, the increasing threshold for Nyquist constraint can make the stopband energy as well as the stopband attenuation improved. By contrast, they can provide FBMC better frequency selectivity than the popular PHYDYAS and IOTA filters. Especially, near the end of transition band, the sidelobes of designed NPR filter are more suppressed than those of the PHYDYAS filter. In the BER tests, the designed NPR filter outperforms the PHYDYAS and IOTA filters under scenarios of additive white Gaussian noise (AWGN) and carrier-frequency offset (CFO), which demonstrates the advantage of the proposed method for NPR prototype filter.

The remaining parts of this paper are organized as follows. Section 2 gives some preliminaries of FBMC and its prototype filter. The proposed NPR design of FBMC prototype filter is presented in Sect. 3. In Sect. 4, some typically designed NPR filters are compared to the PHYDYAS and IOTA filters in BER simulation. Conclusions are drawn in Sect. 5.

2 Preliminaries

2.1 FBMC

The theory of FBMC has been developed over the past five decades, in which the cosine modulated multi-tone (CMT) and the staggered multi-tone (SMT) were

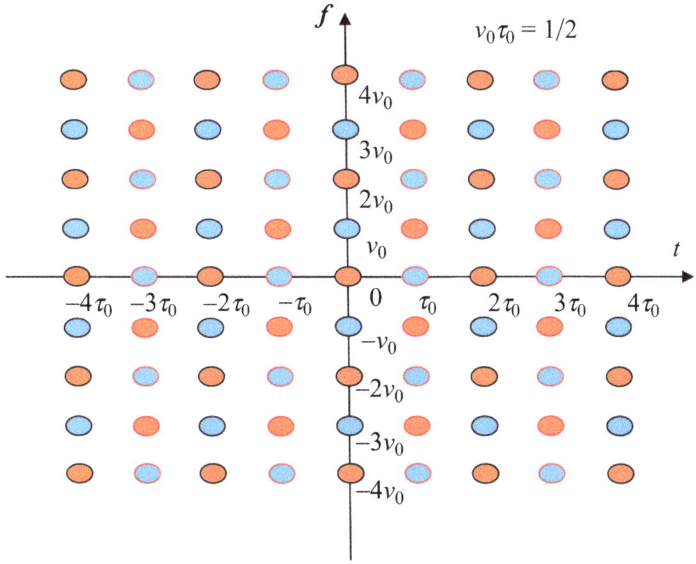

Fig. 1 The time-frequency lattice of SMT

particularly studied for their advantage of bandwidth efficiency. Actually, these two implements of FBMC can be mutually transformed through frequency shift and time-frequency scaling [10]. Thus, the SMT is typically employed in this paper. Figure 1 presents the time-frequency lattice of SMT, in which OQAM operations make the time space half of the symbol duration ($\tau_0 = T_0/2$) and each pair of adjacent symbols along time is treated as the real and imaginary parts of a QAM symbol.

For each constellation of this SMT lattice, its corresponding subcarrier and time can be denoted as $m\upsilon_0$ and $n\tau_0$. Then, the filter for this point can be generated by time-frequency shift from the real-valued prototype filter $g(t)$, i.e., $g(t - n\tau_0)e^{j2\pi m\upsilon_0 t}$. Consequently, the transmitted signal for the real symbol $a_{m,n}$ can be given as below.

$$s(t) = \sum_{m=0}^{M-1} \sum_{n\in\mathbb{Z}} a_{m,n} \underbrace{g(t - n\tau_0)e^{j2\pi m\upsilon_0 t} e^{j\phi_{m,n}}}_{g_{m,n}(t)} \tag{1}$$

where M is the number of FBMC subcarriers and $\phi_{m,n} = (\pi/2)(m+n)$ is the phase term related to the OQAM staggering.

On the receiver, the estimated symbol is acquired after applying a matched filter and the operation of real part taking.

$$\hat{a}_{m_0,n_0} = \text{Re}\left\{\sum_{n\in\mathbb{Z}}\sum_{m=0}^{M-1} a_{m,n}e^{j\phi_1}e^{j\phi_2}A_g((n_0-n)\tau_0,(m_0-m)\upsilon_0)\right\} \tag{2}$$

where $\phi_1 = (\pi/2)(m - m_0 + n - n_0)$ and $\phi_2 = (\pi/2)(m - m_0)(n + n_0)$ Besides, A_g is the ambiguity function of $g(t)$, which is defined as follows [10, 19].

$$A_g(\tau,\nu) = \int_{-\infty}^{+\infty} g(t+\frac{\tau}{2})g(t-\frac{\tau}{2})e^{-j2\pi\nu t}dt \tag{3}$$

According to the expression in (2), it can be deduced that the next equation in (4) must be satisfied to realize the perfect reconstruction of FBMC [4, 13], i.e., $\hat{a}_{m_0,n_0} = a_{m_0,n_0}$. Particularly, it can be equivalent to Nyquist condition when $p = 0$, thus this equation is thought of as the generalized Nyquist condition [12].

$$A_g(2q\tau_0, 2p\upsilon_0) = \begin{cases} 1 & p = 0, q = 0 \\ 0 & p \neq 0, q \neq 0 \end{cases} \tag{4}$$

2.2 Prototype Filter

The generalized Nyquist condition requires the FBMC prototype filter make regular zero crossings both along the time and frequency axises. Such eligible prototype filter is called PR filter for FBMC. However, the design of NPR prototype filter is promoted in practice. Because the interference originated form NPR filter bank is rather small in contrast to the residual interference due to transmission channel, and the NPR filter can enhance FBMC frequency selectivity at the controllable cost of generalized Nyquist condition [4].

In the design of NPR prototype filter, certain amounts of distortion are tolerated in reaching the condition in (4). In other words, the Nyquist condition below is usually relaxed in the design of FBMC prototype filter.

$$r(n) = \begin{cases} 1/M, & n = N-1 \\ 0, & n = N-1 \pm kM, k \neq 0 \\ \text{arbitrary}, & others \end{cases} \tag{5}$$

where the filter autocorrelation $(r(n))$ is the convolution of prototype filters in both transmitter and receiver, i.e., $r(n) = g(n) \otimes g(-n)$ with $\otimes$ denoting the convolution operation. As a result, $r(n)$ is symmetric with an odd length $N_r = 2N - 1$. Furthermore, to analysis the Nyquist distortion of NPR prototype filter, the performance measurement below can be applied [16].

$$ISI = 2\sum_{k=1}^{K} |r(N-1-kM)| \Big/ r(N-1) \tag{6}$$

On the other hand, the sidelobe is more concerned than the zero crossings in frequency-domain. A small sidelobe is essential to high spectral efficiency, which also affects the spectrum sensing of FBMC application in cognitive radio [9]. Therefore, the stopband energy of the prototype filter, which determines the leaked out-of-band interference power, is usually minimized [16, 17]. It can be formulated as

$$\begin{aligned} f(\mathbf{g}) &= \textstyle\int_{\omega_s}^{\pi} |G(\omega, \mathbf{g})|^2 d\omega \\ G(\omega, \mathbf{g}) &= \sum_{n=0}^{N-1} g(n) e^{-j\omega n} \end{aligned} \tag{7}$$

where $G(\omega, \mathbf{g})$ is the frequency response of length-N FBMC prototype filter with $\mathbf{g} = [g(0), g(1), \cdots, g(N-1)]^T$.

To reduce the out-of-band emission, the FBMC prototype filter with narrow band has an increased length, which is K times of the symbol period. This K is also called the overlapping factor in PHYDYAS filters, which determines the number of k in (5). It is generally set as $K = 4$ or a larger one. Besides, $\alpha = 1.0$ is adopted as the roll-off factor rather than traditional $\alpha < 1.0$, because the adjacent orthogonality in lattice is guaranteed by the OQAM staggering [10, 14]. As a result, the starting frequency of filter stopband is $\omega_s = 2\pi / M$, and the original optimization problem considered in literature can be written as

$$\begin{aligned} & f(\mathbf{g}) = \textstyle\int_{2\pi/M}^{\pi} |G(\omega, \mathbf{g})|^2 d\omega \\ & s.t. \begin{cases} 2\sum_{k=1}^{K} \left|\mathbf{g}^T \mathbf{S}_{kM} \mathbf{g}\right| \le ISI^m / M \\ \mathbf{g}^T \mathbf{S}_{KM} \mathbf{g} = 1/M \end{cases} \end{aligned} \tag{8}$$

where $[\mathbf{S}_n]_{p,q} = \begin{cases} 1, & p-q=n \\ 0, & otherwise \end{cases}$ and ISI^m is the threshold to constrain the distortion of Nyquist condition. Increasing the value of ISI^m, lower stopband energy can be achieved because of the implicit performance tradeoff, which can benefit the frequency selectivity of FBMC.

3 Design of the NPR Prototype Filter

Different from the iterative LS method [16] or the SOCP method [17], the proposed design of NPR prototype filter is accomplished by two phases. In the first phase, the cost function and the constrain inequations in (8) are converted to linear ones about

$r(n)$, then the linear programming algorithm is employed to solve this optimization problem. In the second phase, an spectral factorization is employed to retrieve the prototype filter $g(n)$ from its autocorrelation coefficients $r(n)$.

3.1 The Constrained Optimization of $r(n)$

Due to the convolution operation, the Nyquist constraints in (8) are highly nonconvex. To circumvent this problem, the convolution relation between $g(n)$ and $r(n)$ is utilized, then the original inequations are transformed to the linear ones about $r(n)$.

$$\begin{cases} |r(N-1-kM)| \le \mu_k, & kM \ne N-1 \\ |r(kM)| = 1/M, & kM = N-1 \\ \quad 2\sum_{k=1}^{K} \mu_k \le M \cdot ISI^m \end{cases} \tag{9}$$

In this way, the Nyquist condition can be accurately controlled and relaxed for better sidelobe suppression in NPR filter design, as long as $g(n)$ can finally be retrieved from the optimum $r(n)$. The spectral factorization for that retrieval is according to the method in [18].

For $r(n)$ and $g(n)$, apart from the convolution in time-domain, they are also square-related in frequency-domain, i.e., $R(\omega, \mathbf{r}) = |G(\omega, \mathbf{g})|^2$, in which $R(\omega, \mathbf{r})$ is the zero-phase response of $r(n)$ and can be readily expressed by a matrix product.

$$R(\omega, \mathbf{r}) = \mathbf{c}^T(\omega)\mathbf{r} \tag{10}$$

with the definitions

$$\begin{aligned} \mathbf{r} &= [r(0), r(1), \cdots, r(N-1)]^T \\ \mathbf{c}(\omega) &= [2\cos(\omega(N-1)), 2\cos(\omega(N-2)), \cdots, 2\cos(\omega), 1]^T \end{aligned} \tag{11}$$

Accordingly, the original cost function in (7) can be transformed to

$$f(\mathbf{r}) = \int_{2\pi/M}^{\pi} R(\omega, \mathbf{r}) d\omega = \mathbf{s}^T \mathbf{r} \tag{12}$$

where

$$\mathbf{s} = \left[-2\frac{\sin(2\pi/M \cdot (N-1))}{N-1}, \cdots, -2\frac{\sin(2\pi/M)}{1}, \pi - 2\pi/M \right] \tag{13}$$

In addition, the constraints on amplitude ripples are also considered in the subband $[0, \pi/M]$, to make sure the NPR filter nearly fits the 3dB requirement in Nyquist condition.

$$\begin{cases} \mathbf{c}^T(\omega)\mathbf{r} \le (1+\delta_c^{um})^2, & \omega \in [0, \pi/M] \\ \mathbf{c}^T(\omega)\mathbf{r} \ge (1-\delta_c^{lm})^2, & \omega \in [0, \pi/M] \end{cases} \tag{14}$$

where δ_c^{um} and δ_c^{lm} represent respective threshold of upper bound and lower bound.

Finally, combining the cost function (12) with constraint inequations (9) and (14), the design problem for Nyquist filter, i.e., equivalently the autocorrelation coefficients $r(n)$, can be proposed as the following optimization model.

$$\min_{\mathbf{x}} f(\mathbf{x}) = [\mathbf{s}^T, \mathbf{0}_{1\times K}]\mathbf{x}$$
$$s.t. \begin{cases} [\mathbf{c}^T(\omega), \mathbf{0}_{1\times K}]\mathbf{x} \le (1+\delta_c^{um})^2, & \omega \in [0, \pi/M] \\ [\mathbf{c}^T(\omega), \mathbf{0}_{1\times K}]\mathbf{x} \ge (1-\delta_c^{lm})^2, & \omega \in [0, \pi/M] \\ [\mathbf{c}^T(\omega), \mathbf{0}_{1\times K}]\mathbf{x} \ge 0, & \omega \in [0, \pi] \\ [\mathbf{0}_{1\times(N-kM-1)}, 1, \mathbf{0}_{1\times(kM)}, \mathbf{T}_k^1]\mathbf{x} \le 0 \\ 0 \le [\mathbf{0}_{1\times(N-kM-1)}, 1, \mathbf{0}_{1\times(kM)}, \mathbf{T}_k^2]\mathbf{x} \\ [\mathbf{0}_{1\times(N-1)}, 1, \mathbf{0}_{1\times K}]\mathbf{x} = 1/M \\ [\mathbf{0}_{1\times N}, \mathbf{1}_{1\times K}]\mathbf{x} \le M \cdot ISI^m/2 \end{cases} \tag{15}$$

where $\mathbf{x} = [\mathbf{r}^T, \mu_1, \mu_2, \cdots, \mu_K]^T$, and $\mathbf{T}_k^1$ ($\mathbf{T}_k^2$) is a length-K vector with only one non-zero element 1(-1), like $\mathbf{T}_3^1 = [0, 0, -1, 0, \cdots, 0]$. Besides, the third inequation is the necessary and sufficient condition for $r(n)$ to correspond to the autocorrelation of $g(n)$.

Once the filter autocorrelation coefficients $r(n)$ are obtained, a spectral factorization of [18] can be applied to produce the prototype filter $g(n)$. However, the factorization will not be explained in this paper, people can find the details in [18].

3.2 The Example

In this example, the NPR prototype filter with length $N = KM + 1$ ($K = 4$, $M = 32$) is designed by the proposed method. Besides, the threshold ISI^m for relaxed Nyquist condition is taken from $\{10^{-4}, 10^{-1}\}$, and the requirement for the subband amplitude ripples of $r(n)$ is relaxed to 6.5dB.

After optimal solving of the problem (15) and doing spectral factorization, the performances of $r(n)$ and its subfilters $\{\mathbf{g}1, \mathbf{g}2\}$ are totally presented in Table 1. Meanwhile, all their magnitude responses are shown in Fig. 2. Among the performances in Table 1, the definitions for A_s and E_s are given as

$$\begin{cases} R_p = -20\log_{10}(1-\delta_p) & \left(\delta_p = \max\limits_{\omega\in[0,\pi/M]} \left||G(\omega, \mathbf{g})| - 1\right|\right) \\ A_s = -20\log_{10}(\delta_s) & \left(\delta_s = \max\limits_{\omega\in[2\pi/M,\pi]} |G(\omega, \mathbf{g})|\right) \end{cases} \tag{16}$$

Table 1 The performances of the optimized $r(n)$ and the retrieved subfilters $\mathbf{g}1$ and $\mathbf{g}2$ ($K = 4$, $M = 32$)

ISI^m		10^{-4}	10^{-1}
$f(\mathbf{x})$		$8.0768 \cdot 10^{-8}$	$1.0792 \cdot 10^{-8}$
E_s of $r(n)$		$1.3772 \cdot 10^{-13}$	$2.4393 \cdot 10^{-15}$
A_s of $r(n)$		99.6063	116.4661
R_p of $r(n)$		6.0215	6.9357
E_s	$\mathbf{g}1$	$8.1643 \cdot 10^{-8}$	$1.1978 \cdot 10^{-8}$
	$\mathbf{g}2$	$8.0233 \cdot 10^{-8}$	$1.1866 \cdot 10^{-8}$
A_s	$\mathbf{g}1$	49.4854	57.4896
	$\mathbf{g}2$	50.1216	58.1271
R_p	$\mathbf{g}1$	2.9962	3.0347
	$\mathbf{g}2$	3.0259	3.0517
ISI		10^{-4}	10^{-1}

In Table 1, the E_s of the retrieved sub-filters are approaching the cost function $f(\mathbf{x})$, demonstrating not only the effectiveness of the cost function in optimization problem (15) but also the effectiveness of performance inheritance in spectral factorization. Meanwhile, the A_s and R_p of the sub-filters are nearly half about the corresponding A_s and R_p of $r(n)$, which can also be explicitly seen in Fig. 2. This is because the square relation of $R(\omega, \mathbf{r}) = |G(\omega, \mathbf{g})|^2$ has been adopted in establishing the constrained problem in (15). Further, increasing ISI^m from 10^{-4} to 10^{-1}, the more relaxed Nyquist condition can bring both improvements of A_s and E_s. Correspondingly in Fig. 2, the stopband ripples are decreased as a whole and the magnitude responses in passband begin to fluctuate. Therefore, the proposed method based on the spectral factorization in [18] is effective to design NPR filter. It can flexibly realize performance tradeoff for better stopband suppression.

4 Simulation and Analysis

This section presents the comparative study of the design and application of NPR FBMC prototype filter. First, we investigate the influence of ISI relaxation, and do comparison with popular filters, including the PHYDYAS [15] and the IOTA [14]. Second, FBMC applications comprising these NPR filters are tested by computer simulations, where BERs are conducted under scenarios of AWGN and CFO.

4.1 *The Design of NPR Filter*

The performance results of our design, i.e., A_s and E_s, for $M = \{32, 128\}$ and $ISI^m \in \{\{0.01, 0.05\} \cup \{0.2 : 0.4 : 3\} \cup \{4 : 1 : 10\}\} \cdot 10^{-2}$ are displayed in Fig. 3. For distinct showing of E_s, the transformation $-20\log_{10}(E_s)$ is employed.

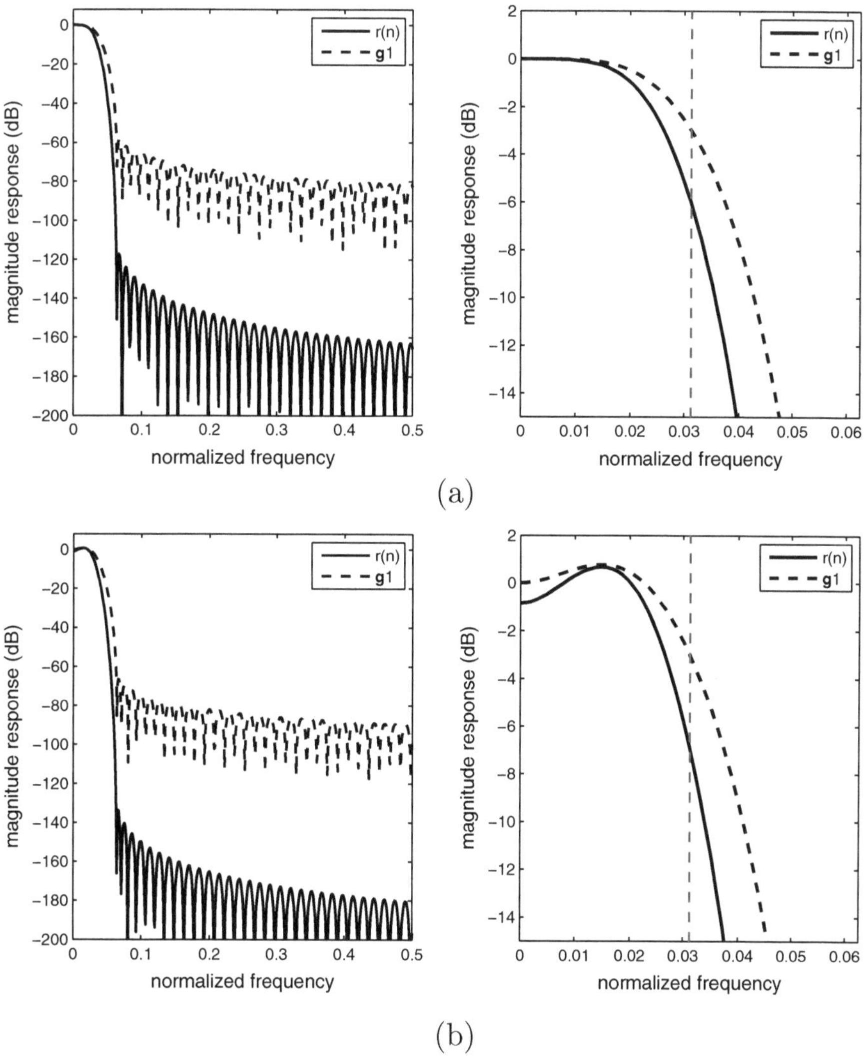

Fig. 2 The magnitude responses of the optimized $r(n)$ and the retrieved sub-filters **g**1 for both cases of $ISI^m = 10^{-4}$ and $ISI^m = 10^{-1}$. (**a**) $ISI^m = 10^{-4}$. (**b**) $ISI^m = 10^{-1}$

Moreover, only the performances of sub-filter **g**1 are presented in this figure due to the approximating performances of the sub-filters **g**1 and **g**2.

From Fig. 3, it is explicit to see the growing of threshold ISI^m, i.e., the relaxing of Nyquist condition, brings improvements of stopband suppression, i.e., stopband attenuation A_s and stopband energy E_s. According to the variation of curves, such improvements are sensitive to small ISI^m. Once ISI^m is larger enough, the improvements greatly slow down and some fluctuations appear, this is why we do

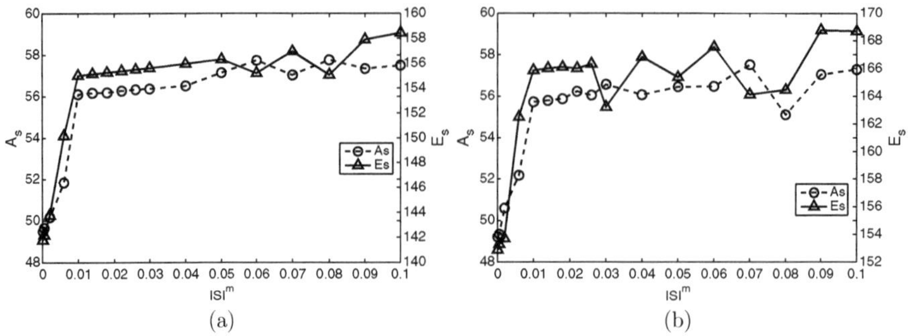

Fig. 3 The tradeoffs of NPR filter performances $\{E_s, A_s\}$ due to the relaxation of ISI^m in the proposed design with $M = 32$ and $M = 128$. (**a**) $M = 32$. (**b**) $M = 128$

Table 2 The performance comparisons of FBMC prototype filters

$\{M = 32, K = 4\}$	PHYDYAS	IOTA	$\mathbf{g1}_A$	$\mathbf{g1}_B$	$\mathbf{g1}_C$
A_s	39.8544	13.5090	49.4854	51.8454	57.4896
E_s	$2.6952 \cdot 10^{-6}$	$1.3125 \cdot 10^{-3}$	$8.1643 \cdot 10^{-8}$	$3.0999 \cdot 10^{-8}$	$1.1978 \cdot 10^{-8}$
ISI	$8.1 \cdot 10^{-4}$	$1.9 \cdot 10^{-2}$	$1.0 \cdot 10^{-4}$	$6.0 \cdot 10^{-3}$	$1.0 \cdot 10^{-1}$
$\{M = 128, K = 4\}$	PHYDYAS	IOTA	$\mathbf{g1}_D$	$\mathbf{g1}_E$	$\mathbf{g1}_F$
A_s	39.8544	13.7545	49.2159	52.1752	57.2801
E_s	$6.7380 \cdot 10^{-7}$	$2.9593 \cdot 10^{-4}$	$2.2687 \cdot 10^{-8}$	$7.5017 \cdot 10^{-9}$	$3.6844 \cdot 10^{-9}$
ISI	$8.1 \cdot 10^{-4}$	$5.2 \cdot 10^{-3}$	$1.0 \cdot 10^{-4}$	$6.0 \cdot 10^{-3}$	$1.0 \cdot 10^{-1}$

not relax ISI^m larger than 10^{-1}. On the other hand, M also affects the performance tradeoff for better E_s and A_s, but its effect is less than that of ISI^m. A larger M can contribute to better E_s while no obvious enhancement of A_s.

To reveal the difference or the advantage of NPR filters designed by the proposed method, the PHYDYAS filter and the IOTA filter are also adopted for comparison. Without loss of generality, only the retrieved prototype filters **g**1 under the Nyquist thresholds $ISI^m = \{10^{-4}, 6.0 \cdot 10^{-3}, 10^{-1}\}$ are taken into account for both cases of $M = 32$ and $M = 128$. The detailed comparisons are shown in next Table 2 and Fig. 4. In the figure, the magnitude responses occupying the width of three subcarriers are displayed for their importance in inter-carrier interference issue.

From Table 2, we can explicitly see the advantages of the NPR filters designed by the proposed method, since they outperform the PHYDYAS filters and the IOTA filters in terms of A_s and E_s. In contrast with the IOTA filters, the proposed filters have an overwhelming superiority in no matter frequency selectivity or sidelobe falloff rate, which is obvious in Fig. 4. Focusing the comparison between the PHYDYAS filters and the designed NPR filters, it can be found that the difference of A_s and E_s can reach up to more than 10dB and two orders of magnitude. In addition, when the controllable ISI is comparative to or smaller than that of the PHYDYAS filter, the designed NPR filter is still superior in stopband suppression. This is because the requirement in the design of PHYDYAS filter, i.e., the requirement of

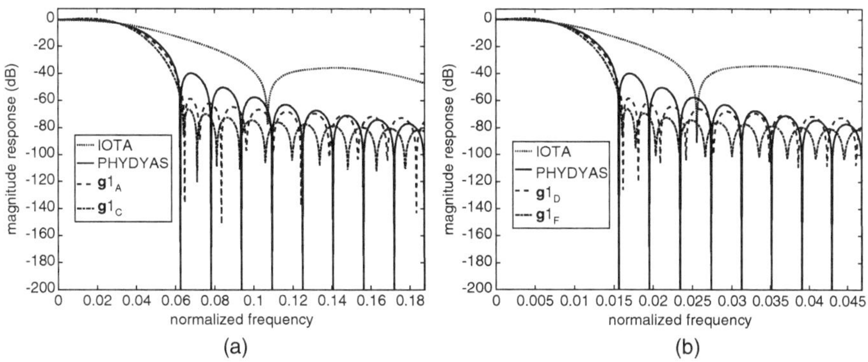

Fig. 4 The magnitude responses of the FBMC prototype filters in Table 2. (**a**) $M = 32$. (**b**) $M = 128$

sidelobe falling rate, do not exit in the proposed method, which can provide more freedom in performance tradeoff. From Fig. 4, the more attenuated sidelobes of the designed NPR filters can be found near the frequency point $2\pi/M$, although the PHYDYAS filters have overall fast fallings of sidelobes. Actually, these sidelobes near the transition, i.e., the frequency response overlapping with several nearby subcarriers, largely determine the inter-carrier interference in FBMC. Therefore, such attenuated sidelobes of the designed NPR filters can benefit the system interference robustness. That will be demonstrated in next BER test of FBMC.

4.2 *The BER Test of FBMC*

Here we present the BER test results for FBMC prototype filters in Table 2. The computer simulations are conducted in the AWGN and CFO scenarios, in which the relative carrier frequency offset (rCFO) is set by two values $\{0, 0.08\}$. The system settings of FBMC are simplified according to those in [4] for $M = 32$ and $M = 128$, and the input signals before OQAM staggering are generated by 4-QAM. After FBMC simulation and BER test, the results are displayed in Fig. 5.

From Fig. 5, we explicitly see that the strong interference in the case of rCFO = 0.08 significantly degrades the BER performances. Meanwhile, the NPR prototype filters designed by the proposed method outperform the PHYDYAS filters and the IOTA filters in BER tests. In AWGN scenario, due to the adjacent orthogonality is approximately guaranteed by the OQAM staggering, the interference for each lattice point in Fig. 1 mainly comes from its neighbouring points, which can be reduced if the sidelobes near the frequency point $2\pi/M$ are suppressed enough. Therefore, all the designed NPR filters with improved A_s and E_s have an advantage on BER, especially the $\mathbf{g}1_{\mathrm{C}}$ and $\mathbf{g}1_{\mathrm{F}}$ with the largest relaxed Nyquist condition. When CFO is added into FBMC simulation, the adjacent orthogonality is destroyed,

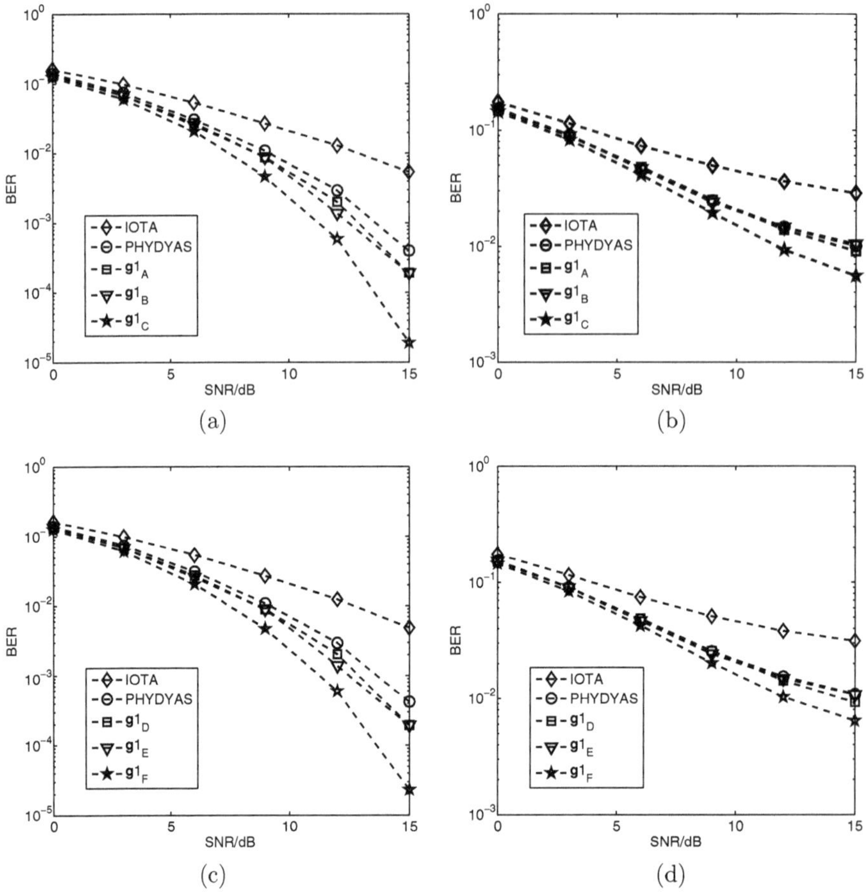

Fig. 5 The BER results of the FBMC simulations, and the filters in Table 2 are employed. (**a**) AWGN ($M = 32$). (**b**) rCFO=0.08 ($M = 32$). (**c**) AWGN ($M = 128$). (**d**) rCFO=0.08 ($M = 128$)

then the advantage attributed by suppressed sidelobe is reduced and the BER gaps between the PHYDYAS filter and the designed NPR filters are diminished. According to above results, it is reasonable to relax the Nyquist condition for better stopband suppression in the design of NPR prototype filter, and it can be efficiently implemented by the proposed method.

5 Conclusions

In this paper, we have achieved a flexible design of FBMC waveform, i.e., the corresponding NPR prototype filter. The stopband suppression is strengthened through the optimization of stopband energy with relaxed Nyquist condition, in which the autocorrelation coefficients are employed to convert the original nonconvex constraints to linear ones and an efficient spectral factorization is used to retrieve the final NPR filter coefficients. From the BER tests in FBMC simulations, the designed NPR filters are demonstrated to be superior to the widely-used PHYDYAS filter and IOTA filter, benefiting the practical applications of FBMC.

Acknowledgement This work is supported by National Natural and Science Foundation of China (61471322).

References

1. Ijaz, A., Zhang, L., Grau, M., Mohamed, A., Vural, S., Quddus, A.U., Imran, M.A., Foh, C.H., Tafazolli, R.: Enabling massive IoT in 5G and beyond systems: PHY radio frame design considerations. IEEE Access **4**, 3322–3339 (2016)
2. Chvez-Santiago, R., Szydelko, M., Kliks, A., Foukalas, F., Haddad, Y., Nolan, K.E., Kelly, M.Y., Masonta, M.T., Balasingham, I.: 5G: the convergence of wireless communications. Wirel. Pers. Commun. **83**(3), 1617–1642 (2015)
3. Cai, Y., Qin, Z., Cui, F., Li, G.Y., McCann, J.A.: Modulation and multiple access for 5G networks. IEEE Commun. Surv. Tutorials **20**(1), 629–646 (2018)
4. Germany, F.E.H., France, A.C.: Final 5GNOW transceiver and frame structure concept D3.3. 5GNOW report (2015). https://www.is-wireless.com/fp7-5gnow/
5. Levy, D., Reichman, A.: Filter bank multi carrier modulation performance. In: 2017 IEEE International Conference on Microwaves, Antennas, Communications and Electronic Systems (COMCAS), pp. 1–6. IEEE, Israel (2017)
6. Zhao, Z., Gong, X., Schellmann, M.: A novel FBMC/OQAM scheme facilitating MIMO FDMA without the need for guard bands. In: 20th International ITG Workshop on Smart Antennas (WSA 2016), pp. 1–5. VDE, Germany (2016)
7. Aminjavaheri, A., Farhang, A., RezazadehReyhani, A., Farhang-Boroujeny, B.: Impact of timing and frequency offsets on multicarrier waveform candidates for 5G. In: 2015 IEEE Signal Processing and Signal Processing Education Workshop (SP/SPE), pp. 178–183. IEEE, United states (2015)
8. Dor, J.B., Berg, V., Ktnas, D.: Channel estimation techniques for 5G cellular networks: FBMC and multiuser asynchronous fragmented spectrum scenario. Trans. Emerg. Telecommun. Technol. **26**(1), 15–30 (2015)
9. Hosseini, H., Anpalagan, A., Raahemifar, K., Erkucuk, S., Habib, S.: Joint wavelet-based spectrum sensing and FBMC modulation for cognitive mmWave small cell networks. IET Commun. **10**(14), 1803–1809 (2016)
10. Farhang-Boroujeny, B.: OFDM versus filter bank multicarrier. Signal Process. Mag. IEEE **28**(3), 92–112 (2011)
11. Farhang-Boroujeny, B.: Filter bank multicarrier modulation: a waveform candidate for 5G and beyond. Adv. Electr. Eng. **2014**, Article ID 482805, 25 (2014). https://doi.org/10.1155/2014/482805
12. Sahin, A., Guvenc, I., Arslan, H.: A survey on multicarrier communications: prototype filters, lattice structures, and implementation aspects. IEEE Commun. Surv. Tutorials **16**(3), 1312–1338 (2014)
13. Shaeen, K., Elias, E.: Prototype filter design approaches for near perfect reconstruction cosine modulated filter banks-a review. J. Signal Process. Syst. **81**(2), 183–195 (2015)
14. Sahin, A., Guvenc, I., Arslan, H.: A comparative study of FBMC prototype filters in doubly dispersive channels. In: 2012 IEEE Globecom Workshops (GC Wkshps), pp. 197–203. IEEE, United states (2012)

15. Aminjavaheri, A., Farhang, A., Doyle, L.E., Farhang-Boroujeny, B.: Prototype filter design for FBMC in massive MIMO channels. In: 2017 IEEE International Conference on Communications (ICC), pp. 1–6. IEEE, France (2017)
16. Farhang-Boroujeny, B.: A square-root nyquist (M) filter design for digital communication systems. IEEE Trans. Signal Process. **56**(5), 2127–2132 (2008)
17. Lu, W.S., Saramaki, T., Bregovic, R.: Design of practically perfect-reconstruction cosine-modulated filter banks: a second-order cone programming approach. IEEE Trans. Circuits Syst. I Reg. Pap. **51**(3), 552–563 (2004).
18. Hua, J., Wen, J., Lu, W., Li, F., Jiang, B.: Design and application of nearly Nyquist and SR-Nyquist FIR filter based on linear programming and spectrum factorization. In: 9th Conference on Industrial Electronics and Applications (ICIEA), pp. 64–67. IEEE, China (2014)
19. Choi, J.M., Oh, Y., Lee, H., Seo, J.S.: Pilot-aided channel estimation utilizing intrinsic interference for FBMC/OQAM systems. IEEE Trans. Broadcast. **63**(4), 644–655 (2017)

Robust Spectrum-Energy Efficiency for Green Cognitive Communications

Cuimei Cui, Dezhi Yang, and Shi Jin

1 Introduction

Cognitive radio technologies are envisaged to alleviate the problems of both inefficient utilization and relative scarcity of spectrum by opportunistically exploiting the existing licensed spectrum. However, due to the growth of low power consumption devices in massive Internet of Things and 5G communication environments, energy efficiency also faces the serious challenges and attracts the researchers more attention. Therefore, in a wireless communication system, two major challenges are to achieve high spectral and energy efficiency.

To enhance both spectrum and energy efficiencies of the wireless communication system, so that to achieve green cognitive communication in the future wireless networks, one of the possible solution for these two challenges is radio frequency (RF) energy harvesting cognitive radio network (EH-CRN) [1]. For this EH-CRN technique, it allows cognitive users to harvest electromagnetic waves from ambient radio frequency sources, and convert them into power energy which can be used for wireless transmission. In cognitive radio networks (CRNs), the radio frequency signals transmitted by the licensed users (also named as primary users, PUs) would be no longer give rise to interference for the cognitive radio users (also named as secondary users, SUs) provided that the capability of energy harvesting is integrated

C. Cui (✉)
Southeast University, Nanjing, China

Changzhou Institute of Technology, Changzhou, China

D. Yang
Soochow University, Suzhou, China

S. Jin
Southeast University, Nanjing, China

B. Ye et al. (eds.), *2nd International Conference on 5G for Ubiquitous Connectivity*, EAI/Springer Innovations in Communication and Computing,
https://doi.org/10.1007/978-3-030-22316-8_10

into SUs, but can be identified as a green energy harvesting resources. By taking advantage of this technology, SUs can get the utmost out of both the spectrum and energy of PUs. As for the state-of-art of EH-CRNs research, they have been studied operating in interweave mode in [2–4], overlay mode in [5–7] and underlay mode in [7–10].

In interweave mode EH-CRNs, SUs first harvest RF energy from ambient PUs, and then sense the state of PUs and opportunistically access the licensed spectrum if PUs are inactive. In [2], the selection problem of dynamic channels in a multi-channel RF-powered CRN was discussed to obtain the optimal channel by formulating a Markov decision process problem. Then, a dynamic centralized cooperative spectrum sensing scheme and its corresponding access policy were proposed to achieve the maximized throughput with the available energy for EH-CRNs in [3]. In [4], under energy causality and collision constraints, the authors proposed an optimal spectrum sensing policy to maximize the expected total throughput of SUs in EH-CRNs.

In overlay mode EH-CRNs, SUs supply RF energy to both PUs and SUs on condition that there are harmonious collaboration between PUs and SUs. On this aspects, the maximization of sum-throughput is studied in [5], wherein a hybrid access point first transfers wireless energy for a number of SUs and supplies information for PUs, and then collects sensed data from SUs. Similarly, the maximization of energy efficiency was discussed in [6], at the same time, the uplink scheduling and cooperative power control were also considered. In [7], a hybrid underlay-overlay EH-CRNs was modeled, and also the access strategy combined with the framework of partially observable Markov decision process was proposed to maximize the system long-term throughput.

In underlay mode EH-CRNs, SUs work utilizing their harvested RF energy until the interference to PUs exceed a tolerable threshold value. In [8], an efficient strategy of relay selection was designed to determine when and which relay should be selected to minimize outage transmission probability in EH-CRNs with spatial randomly distributed SUs. Moreover, the maximization problem of end-to-end throughput in terms of optimal time and power allocation is studied for multi-hop EH-CRNs in [9]. Recently, an optimal offline harvest-or-transmit strategy is proposed in [10] to maximize the achievable rate of SU under energy causality and interference constraints for an underlay EH-CRN which operates in slotted fashion.

Based on the comparative analysis of the above, we can observe that these studies almost separately focus on various problems. In contrast, this paper is to study a green cognitive communications network, where PUs coexist with SUs, to enhance spectrum efficiency and green energy utilization efficiency simultaneously. To be specific, we set up an underlay EH-CRN with multi antennas and multiple battery-free SUs to lower transmit powers of SUs. Based on our previous works, we further design a new EH-CRN model which operates in the dynamic and variable scheduling mechanism of time division multiple access (DV-TDMA) [11], each SU first senses whether PU is active or inactive, then harvests energy from the RF signals energy of PU transmitter if PU is in active, or else, transmits data in the allocated time sequentially or keep idle.

The rest of this paper is organized as follows. In Sect. 2, we describe a heterogeneous model of EH-CRN with 5G. An optimal energy efficiency scheme is proposed to achieve the maximization of the system throughput by discussing the optimal power splitting scheme in Sect. 3. Then the simulation analysis and performance evaluation is provide in Sect. 4. Finally, the comprehensive conclusions are drawn in Sect. 5.

2 System Model

We consider a multichannel EH-CRN architecture in 5G communications scenarios, as shown in Fig. 1, where primary user network coexists with cognitive radio network. Femtocell is assumed as cognitive radio network with multichannel, it consists of K cognitive radio users (act as secondary users Transmitter, SU-TXs) and one cognitive radio base with M antennas (acts as secondary users receiver, SU-RX), $M \geq K$. Picocell is assumed as primary user network, it consists of N PU receivers (named as PU-RXs) and one PU Transmitter (named as PU-TX) with L antennas. Both PU-RXs and SU-TXs are set up with one single antenna, the nth PU-RX and k^{th} SU-TX indexed by PU-RX_n and SU-TX_k, respectively, and $n \in \text{N} \triangleq \{1, \ldots, N\}$, $k \in \kappa \triangleq \{1, \ldots, K\}$. The SU-RX employs zero-forcing receiver to sense the multiple simultaneous signals transmitted by the kth SU-TXs at the same frequency band. In addition, each of SU-TXs first estimates its received power energy from PU-TXs to decide to adopt the individual sensing or cooperative spectrum sensing scheme corresponding to the strong or weak signal transmitted by PU-TXs. The data fusion center SU-RX collects all detected results and the cross information between PU-TX and SU-RX for cooperative spectrum sensing and centralized power control. Here, we assume that all SUs employ a separate control channel to avoid overlapping with PUs' channels when the control signals, such as sensing results and the channel information, are in the process of transmission and reception.

In the proposed EH-CRN architecture, we consider all PUs operate in time division duplexing mode with slot duration T, each of PUs alternates in ON/OFF switch channels, their busy and idle durations follow independent and identically distributed exponential distributions with mean $1/\lambda_0$ and $1/\lambda_1$. The busy and idle status of PUs at n^{th} channel (PU-RX_n) are denoted by $I_1^n = 1$ and $I_0^n = 0$, respectively. The CRN is synchronized with one primary user network. The time framework of the CRN adopting the DV-TDMA scheduling fashion in [11], which includes two phases: (1) a cooperative spectrum sensing phase with duration t_s and (2) a data transmission phase with duration $T_a - t_s$ or energy harvesting phase with duration $T_h - t_s$.The uncertainty of the available energy for SUs are obtained by the active probability p_a, which means is that SUs have sufficient energy to finish their task during the period of data transmission provided that PU is sensed to be inactive. In one time frame duration, any of SUs is to sense the status of PUs if it has enough energy to perform spectrum sensing and data transmission, or remains idle otherwise. According to the sensing result, SUs may transmit data if the PU

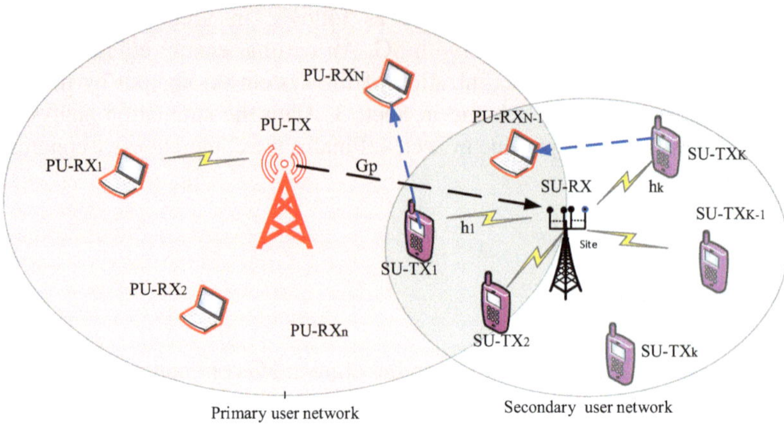

Fig. 1 5G Heterogeneous Cognitive Radio Network Model

detected is in OFF state, or it switches into energy harvesting mode if PU is in ON state. Similarly, the SU also enters the energy harvesting mode when it runs out of energy.

The designed spectrum sensing time framework describes that SU-TXs sense PUs are in ON state or OFF state with sensing time t_s within one time frame duration $T = [T_a, T_h]$ as shown in Fig. 2, where T_a stands for one sensing frame duration and T_h stands for one radio energy harvesting frame duration, and t_s follows the dynamic and variable time division multiple access in [11]. The Fig. 2 depicts two kinds of sequences for PU in ON/OFF periods, as well the corresponding operations of SU. In the first instance, the time slots of SU consists of the sensing time t_s and f a data transmission slot $T_a - t_s$ when SU senses an inactive PU. However, due to the unslotted PU model, the PU can unexpectedly return to its channel that allowed the SU to transmit, this may result in a collision between the two transmissions. Therefore, SU-TX should stop data transmission avoiding the conflict with the PU, and switch to radio energy harvesting with a harvesting slot duration $T_h - t_s$ if the channel of PU found busy as shown in the second instance. But, when the PU is idle during the energy harvesting period of SU, the SU cannot harvest radio frequency energy.

The detection probability $p_{11}^{(n)}$ and the false alarm probability $p_{01}^{(n)}$ for the SU-TX$_k$ at the nth channel, under the dual collaborative spectrum sensing (DCS) scheme with energy detection if it first estimates its received weak power from PU-TX, can be obtained as follows [11],

$$p_{11}^{(n)} = \sum_{x=l}^{K} \binom{K}{x} \left(p_{c,k}^{(n)}\right)^x \left(1 - p_{c,k}^{(n)}\right)^{K-x} \tag{1}$$

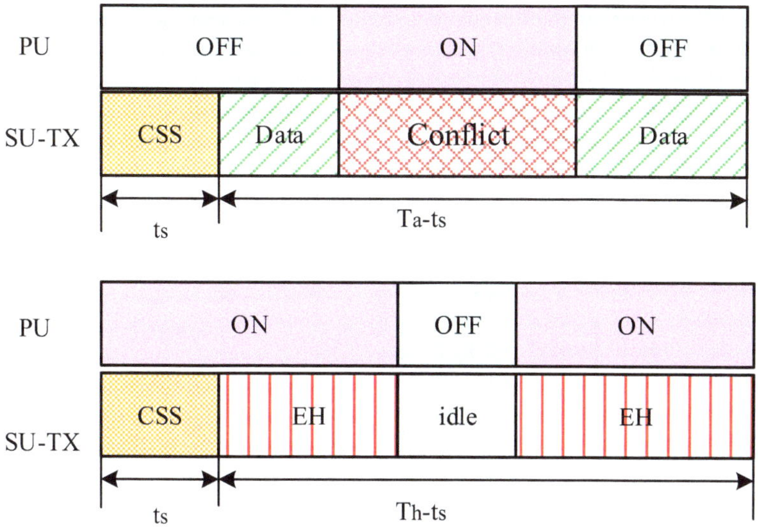

Fig. 2 Time structure of spectrum sensing and energy harvesting

$$p_{01}^{(n)} = \sum_{x=l}^{K} \binom{K}{x} \left(p_{f,k}^{(n)}\right)^{x} \left(1 - p_{f,k}^{(n)}\right)^{K-x} \tag{2}$$

where $p_{c,k}^{(n)}$, $p_{f,k}^{(n)}$ denotes the detection probability and f the false alarm probability for the SU-TX$_k$ at the nth channel under the single cooperative sensing (SCS) scheme. Then, according to the sensing results $\hat{I}_j, j \in \mathrm{B} \triangleq \{0, 1\}$, SU-RX decided the transmitted power $P_{k,j}^{(n)}$ for the SU-TX$_k$ at the nth channel.

According to the renewal process theory, when the PU channel is actually in OFF state during a transmission slot of the SU, if it is sensed to be idle at the beginning of the slot, the corresponding average duration T_{00} can be written as [12],

$$T_{00}\left(T_a\right) = T_a - t_s - u\left(T_a - t_s - \frac{1 - \exp\left[-\left(\lambda_0 + \lambda_1\right)\left(T_a - t_s\right)\right]}{\lambda_0 + \lambda_1}\right) \tag{3}$$

where $u = \lambda_0/(\lambda_0 + \lambda_1)$ is the channel utility of PU in ON state. Similar to T_{11}, when the PU channel is actually in ON state during an energy harvesting slot of the SU, if it was busy at the beginning of the slot, then the corresponding average duration T_{11} is given by,

$$T_{11}\left(T_h\right) = T_h - t_s - u\left(T_h - t_s - \frac{1 - \exp\left[-\left(\lambda_0 + \lambda_1\right)\left(T_h - t_s\right)\right]}{\lambda_0 + \lambda_1}\right) \tag{4}$$

The average duration T_{10} means that the PU channel is busy but it is sensed to be idle, its corresponding formulation is $T_{10}(T_a) = T_a - t_s - T_{11}(T_s)$, and the average duration T_{01} for which the PU channel is idle, but the detection results is in busy, its corresponding formulation is $T_{01}(T_h) = T_h - t_s - T_{00}(T_h)$.

3 Energy Efficiency Optimization

During the process of data transmission, according to the sensing results $\hat{I}_j$, SU-RX received the $M \times 1$ signal vectors [12],

$$y_r = \begin{cases} \sum_{k\in\mathcal{K}} h_k \sqrt{P_{k,j}} x_{s,k} + w_r, & when\ I_0 \\ \sum_{k\in\mathcal{K}} h_k \sqrt{P_{k,j}} x_{s,k} + G_p q x_p + w_r, & when\ I_1 \end{cases} \tag{5}$$

where h_k denotes the instantaneous channel gain between SU-TX$_\mathrm{k}$ and SU-RX (i.e. the kth SU-TX$_\mathrm{s}$ and the secondary user receiver), G_p denotes the instantaneous channel gain between the primary user transmitter (PU-TX) and the secondary user receiver SU-RX, the normalized beamforming vector at the primary user transmitter PU-TX is denoted as q, $x_{s,\,k} \sim CN(0, 1)$ stands for Gaussian random signal transmitted by the SU-TX$_\mathrm{k}$, $w_r \sim CN\left(0, \sigma_r^2 \mathrm{I}\right)$ is Gaussian random noise received by SU-RX. Assumed $U = (H^H H)^{-1} H^H$, $H = [h_1, \ldots, h_K]$ are $M \times K$ channel gain matrix between the SU-TX$_\mathrm{s}$ and the SU-RX, according to the pairs of sensing results $\left(I_i, \hat{I}_j\right)$, $i, j \in \mathrm{B} \triangleq \{0, 1\}$, the achievable rate of the kth SU-TX$_\mathrm{s}$ at the nth channel is given by [13],

$$\begin{aligned} \left(I_0, \hat{I}_0\right) &: \ r_{k,00}\left(P_{k,0}^{(n)}\right) = \rho_{00}\log_2\left(1 + \gamma_{k,0}^{(n)} P_{k,0}^{(n)}\right) \\ \left(I_1, \hat{I}_0\right) &: \ r_{k,10}\left(P_{k,0}^{(n)}\right) = \rho_{10}\log_2\left(1 + \gamma_{k,1}^{(n)} P_{k,0}^{(n)}\right) \\ \left(I_0, \hat{I}_1\right) &: \ r_{k,01}\left(P_{k,1}^{(n)}\right) = \rho_{01}\log_2\left(1 + \gamma_{k,0}^{(n)} P_{k,1}^{(n)}\right) \\ \left(I_1, \hat{I}_1\right) &: \ r_{k,11}\left(P_{k,1}^{(n)}\right) = \rho_{11}\log_2\left(1 + \gamma_{k,1}^{(n)} P_{k,1}^{(n)}\right) \end{aligned} \tag{6}$$

where $\rho_{00} = T_{00}/T$, $\rho_{10} = T_{10}/T$, $\rho_{01} = 1 - T_{01}/T$, $\rho_{11} = 1 - T_{11}/T$, $\gamma_{k,0}^{(n)} = \left(\left[\sigma_r^2 U U^H\right]_{k,k}\right)^{-1}$ and $\gamma_{k,1}^{(n)} = \left(\left[U\left(\sigma_p^2 G_p q q^H G_p^H + \sigma_r^2 I\right) U^H\right]_{k,k}\right)^{-1}$ are effective channel power gain for the kth SU-TX$_\mathrm{s}$ under I_0 and I_1, respectively. Considering all possible combinations of pair $\left(I_i, \hat{I}_j\right)$, the average achievable throughput of the kth SU-TX$_\mathrm{s}$ at the nth channel is,

$$\overline{r}_k^{(n)}(P_k) = \sum_{i\in B}\sum_{j\in B} p_a p_{ij} r_{k,ij}\left(P_{k,j}^{(n)}\right) \tag{7}$$

where $P_k = \left[P_{k,0}^{(n)}, P_{k,1}^{(n)}\right]$ are the transmitted power matrix vector of the kth SU-TX$_\mathrm{s}$ (SU-TX$_\mathrm{k}$) at the nth channel, and $p_{ij} = \Pr\left[\hat{I}_j | I_i\right] \cdot \Pr[I_i]$ is the probability of

$\left(I_i, \hat{I}_j\right)$. From above, then the average energy efficiency function of the SU-TX$_{\mathrm{k}}$ at the nth channel at the nth channel can be expressed as,

$$\eta_k^{(n)}\left(\mathrm{P_k}\right) = \sum_{i \in B} \sum_{j \in B} p_a p_{ij} \eta_{k,ij}\left(P_{k,j}^{(n)}\right) \tag{8}$$

For $k \in \kappa$, where

$$\eta_{k,ij}\left(P_{k,j}^{(n)}\right) = \frac{r_{k,ij}\left(P_{k,j}^{(n)}\right)}{\rho_{ij} P_{k,j}^{(n)} + P_{k,\mathrm{c}}^{(n)}} \tag{9}$$

is the instantaneous energy efficiency of the SU-TX$_{\mathrm{k}}$ under $\left(I_i, \hat{I}_j\right)$, and $P_{k,\mathrm{c}}^{(n)}$ is the circuit power consumed by SU-TX$_{\mathrm{k.}}$ It should be noted that, we only consider users' energy efficiency rather than the system's energy efficiency because each individual secondary user transmitter SU-TX gets its own throughput $r_{k,ij}\left(P_{k,j}^{(n)}\right)$, but at the cost of its own power consumption $(\rho_{ij} P_{k,j}^{(n)} + P_{k,\mathrm{c}}^{(n)})$, for $k \in \kappa$, $i, j \in \mathrm{B} \triangleq \{0, 1\}$.

The optimized objective function is given by,

$$\max_{\{P_k\}_{k=1}^{K}} \eta \triangleq \left[\eta_1\left(P_1\right), \ldots, \eta_K\left(P_K\right)\right] \tag{10}$$

Subject to,

$$\begin{aligned}
&C_1 : p_a \left[\left(\rho_{00} p_{00} + \rho_{10} p_{10}\right) P_{k,0}^{(n)} + \left(\rho_{01} p_{01} + \rho_{11} p_{11}\right) P_{k,1}^{(n)}\right] \le P_{\max,k}, k \in \kappa \\
&C_2 : p_a \sum_{k \in \kappa} \left|g_{k,n}\right|^2 \left(\rho_{10} p_{10} P_{k,0}^{(n)} + \rho_{11} p_{11} P_{k,1}^{(n)}\right) \le PI_{\max,k}, n \in \mathrm{N} \\
&C_3 : 1 - p_{11}^{(n)} + p_{01}^{(n)} \le \varepsilon, \quad n \in \mathrm{N}
\end{aligned} \tag{11}$$

where $P_{\max,\,k}$ and $PI_{\max,\,k}$ denote the maximum transmitted power and the maximum inference power, $g_{k,\,n}$ stands for instantaneous channel gain between the SU-TX$_{\mathrm{k}}$ and PU-RX$_{\mathrm{n}}$, and ε denotes sensing error rate allowed by the communication system.

4 Simulation Analysis and Evaluation

In this section, we evaluate the performance of our solution by experimental simulation. In this simulation, we assume the active and idle probability of PU

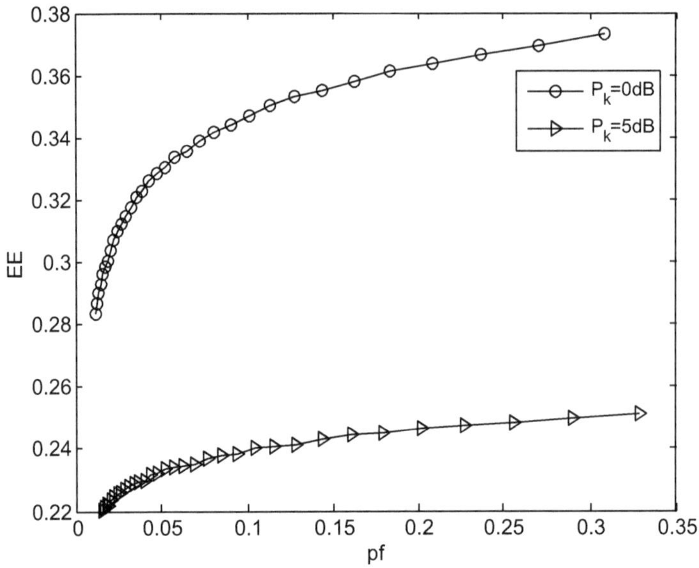

Fig. 3 Energy efficiency (EE) vs. false alarm probability (pf) for different transmitted power

in each channel is $p\left(\mathrm{H}_1^n\right) = p\left(\mathrm{H}_0^n\right) = p_a = 0.5$, $T = 100ms$, $\varepsilon = 0.01$, $P_{\max,k} = 10dB$.

Figure 3 depicts the curves of optimal energy efficiency in terms of false alarm probability. It is clearly observed that the energy efficiency for $P_K = 0dB$ outperform that of $P_K = 5dB$, it's suggest that the energy efficiency is more robust for the lower transmitted power. At another figure, shown as Fig. 4, the energy efficiency of proposed robustness solution outperforms that of no optimization, while meeting the requirement of sensing error rate $\varepsilon \leq 0.01$.

5 Conclusion

In this paper, we study a green cognitive communication networks which PUs coexist with SUs to enhance spectrum efficiency and energy utilization efficiency simultaneously. An opportunistic RF energy harvesting capability of SU is incorporated into sensing time framework, the transmission and RF energy harvesting durations have been obtained based on the traffic pattern of PU. Further, both spectrum and energy efficiency function with respect to transmission power, sensing time, and channel status are formulated. The simulation analysis show that the higher spectrum and energy efficiency can be attained as compared with another schemes.

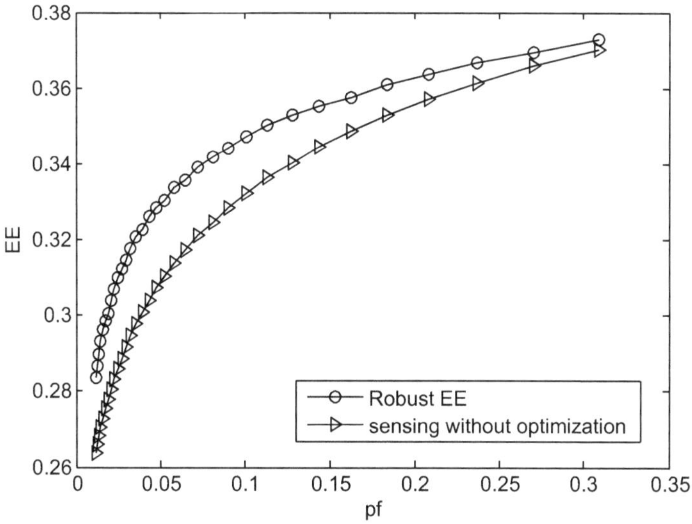

Fig. 4 Robust EE scheme vs. no optimization EE scheme

Acknowledgments This work was supported by National Natural Science Fund of China with Grant No. 61801056, Project Founded by China Postdoctoral Science Foundation with Grant No. 2018M632203, Open fund for Key Laboratory of Agricultural Remote Sensing of Ministry of Agriculture with Grant No. 2017001, Open fund for Jiangsu Key Laboratory of Wireless Communications with Grant No. 2017WICOM07, as well as Technology support project (social development) of Changzhou City with Grant No. CE20185041.

References

1. Huang, X., Han, T., Ansari, N.: On green-energy-powered cognitive radio networks. IEEE Commun. Surv. Tutorials. **17**, 827–842 (2015)
2. Park, S., Kim, H., Hong, D.: Cognitive radio networks with energy harvesting. IEEE Trans. Wirel. Commun. **12**(3), 1386–1397 (2013)
3. Pratibha, Li, K.H., Teh, K.C.: Dynamic cooperative sensing-access policy for energy-harvesting cognitive radio systems. IEEE Trans. Veh. Technol. **65**(12), 10137–11014 (2016)
4. Pratibha, Li, K.H., Teh, K.C.: Optimal spectrum access and energy supply for cognitive radio systems with opportunistic RF energy harvesting. IEEE Trans. Veh. Technol. **66**(8), 7114–7122 (2017)
5. Lee, S., Zhang, R.: Cognitive wireless powered network: spectrum sharing models and throughput maximization. IEEE Trans. Cogn. Commun. Netw. **1**(3), 335–346 (2015)
6. Yin, S., Qu, Z., Wang, Z., Li, L.: Energy-efficient cooperation in cognitive wireless powered networks. IEEE Commun. Lett. **21**(1), 128–131 (2017)
7. Usman, M., Koo, I.: Access strategy for hybrid underlay-overlay cognitive radios with energy harvesting. IEEE Sensors J. **14**(9), 3164–3173 (2014)
8. Yan, Z., Chen, S., Zhang, X., Liu, H.: Outage performance analysis of wireless energy harvesting relay-assisted random underlay cognitive networks. IEEE Internet Things J. **5**(4), 2691–2699 (2018)

9. Xu, C., Zheng, M., Liang, W., Yu, H.B., Liang, Y.C.: End-to-end throughput maximization for underlay multi-hop cognitive radio networks with RF energy harvesting. IEEE Trans. Wirel. Commun. **16**(6), 3561–3572 (2017)
10. Kalpant, P., Adrish, B.: Optimal harvest-or-transmit strategy for energy harvesting underlay cognitive radio network. https://arxiv.org. Last accessed 5 May 2018
11. Cui, C., Wang, Y.: Analysis and optimization of sensing reliability for relay-based dual-stage collaborative spectrum sensing in cognitive radio networks. Wirel. Pers. Commun. **72**(4), 2321–2337 (2013)
12. Park, H., Hwang, T.: Energy-efficient power control of cognitive femto users for 5G communications. IEEE J. Sel. Areas Commun. **34**(4), 772–785 (2016)
13. Cui, C., Yang, D.: Throughput optimization for dual collaborative spectrum sensing with dynamic scheduling. Mod. Phys. Lett. B. **31**(19–21), 1740089-1-6 (2017)

Optimal Precoding Design for LoS Massive MIMO Channels with the Spherical-Wave Model

Lei Yang, Xumin Pu, Shi Jin, Rong Chai, and Qianbin Chen

1 Introduction

The capacity of the LoS MIMO channels with the SWM can be estimated correctly, and it has received much attention [1–6]. By simulations with the SWM, Jiang and Ingram [1] investigated the effects of the transceiver distance, direction angle and elevation angle on the LoS channel capacity. In the three-dimensional geometric model, the optimal antenna spacing for the LoS MIMO channels with the SWM was obtained by maximizing the channel capacity [2]. In [3], the LoS MIMO capacity of the ULAs was studied at the center frequency of 5.2 GHz, and simulation results show that the capacity varies with the locations of transceiver and array element spacing. For the 2×2 3D MIMO channels, the optimal antenna placement in the presence of the path loss and phase differences was studied in [4]. By simulations with the SWM and the PWM, Tamaddondar and Noori [5] analyzed the capacity of the LoS MIMO channels with the non-uniform linear array of antennas.

However, Jiang and Ingram [1], Sarris and Nix [2], Skentos et al. [3], Pu et al. [4], Tamaddondar and Noori [5] are essentially based on the simulations and channel measurements to study the capacity of the LoS channels with the SWM, without discussion of the precoding design. Recently, Pu et al. [6] investigated the transmission scheme for the ULA-based LoS MIMO channels with the SWM and derived the

L. Yang (✉) · X. Pu · R. Chai · Q. Chen
School of Communication and Information Engineering, Chongqing University of Posts and Telecommunications, Chongqing, China
e-mail: s160131185@stu.cqupt.edu.cn; puxm@cqupt.edu.cn; chairong@cqupt.edu.cn; chenqb@cqupt.edu.cn

S. Jin
National Mobile Communications Research Laboratory, Southeast University, Nanjing, China
e-mail: jinshi@seu.edu.cn

B. Ye et al. (eds.), *2nd International Conference on 5G for Ubiquitous Connectivity*, EAI/Springer Innovations in Communication and Computing,
https://doi.org/10.1007/978-3-030-22316-8_11

analytical expressions for the transmission direction and power allocation. However, UCAs have the advantages of compact structure, lower mutual coupling between array elements, and easier use of azimuth and elevation information when compared to ULAs [7]. The main work of this paper is to analyze the precoding for the UCA-based LoS MIMO channels with the SWM. We consider the $4 \times N$ LoS MIMO channels with the ULAs at the Tx and UCAs at the Rx. The corresponding analytical expressions for the precoding matrix and power allocation are obtained by maximizing the channel capacity. The transceiver distance, wavelength, array element spacing and the number of receive antennas are the decisive roles of the precoding design. The simulation results show that the better performance of the proposed scheme is obtained compared with that of the equal power allocation scheme.

The remainder of the paper is organized as follows. Section 2 describes the system model. Section 3 analyzes the low complexity LoS MIMO precoding system with the SWM. Numerical results are provided in Sect. 4. Section 5 is the conclusion of this paper.

2 System Model

In this paper, we consider a single-user communication scenario with the ULAs at the Tx and the UCAs at the Rx, as shown in Fig. 1. We assume that the Tx array is located on the positive half of the x-axis and the Rx array is parallel to the x-y plane. Meanwhile, the center of the circle is located on the positive half of the z-axis. And the distance between the center of UCAs and the origin is D. We further assume that the first receiver antenna element is on the positive half of the x'-axis. In Fig. 1, θ_t $\left(0 \leq \theta_t \leq \frac{\pi}{2}\right)$ denotes the direction angle of the Tx. In addition, R is the radius of the UCAs at Rx, and d_t represents the element spacing of the ULAs at Tx.

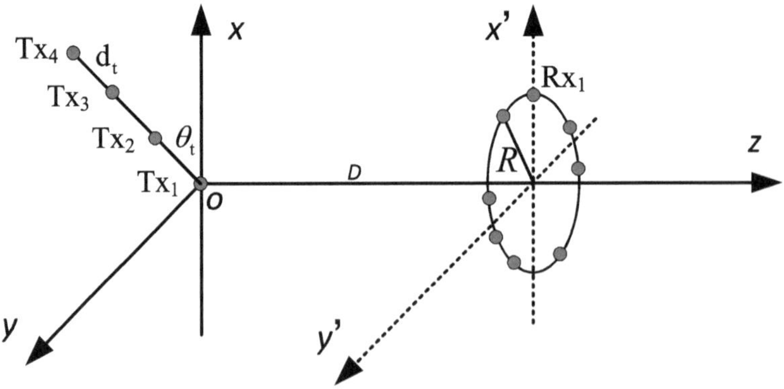

Fig. 1 The antennas arrays for the LoS MIMO channels

Considering a frequency-flat fading channel, the baseband received signal vector can be written as [8]

$$\mathbf{y}{=}\mathbf{H}\mathbf{s}{+}\mathbf{n}, \tag{1}$$

where $\mathbf{H} \in \mathbb{C}^{N\times 4}$ is the channel matrix, $\mathbf{s} \in \mathbb{C}^{4\times 1}$ and $\mathbf{y} \in \mathbb{C}^{N\times 1}$ are, respectively, the transmitted and received signal vector, $\mathbf{n} \in \mathbb{C}^{N\times 1}$ is a complex additive white Gaussian noise (AWGN) vector and $E\left\{\mathbf{n}\mathbf{n}^H\right\} = \sigma^2\mathbf{I}_N$. The ergodic capacity of the system can be written as [8]

$$C = \max_{\mathrm{Tr}(\mathbf{Q})=P} E\left[\log_2 \det\left(\mathbf{I}_N + \frac{1}{\sigma^2}\mathbf{H}\mathbf{Q}\mathbf{H}^H\right)\right], \tag{2}$$

where $\mathbf{Q}{=}E\left\{\mathbf{s}\mathbf{s}^H\right\}$. The power allocated to each eigenvector is given by the eigenvalues of the transmit covariance matrix $\mathbf{Q}$ which satisfying the total power constraint $\mathrm{tr}\left(\mathbf{Q}\right) = P$. Here, $\rho{=}P/\sigma^2$ is defined as the signal-to-noise ratio (SNR). In this paper, we assume that the Tx know the perfect channel state information (CSI). Using the eigenvalue decomposition $\mathbf{Q}{=}\mathbf{U}_{\mathrm{Q}}\Lambda\mathbf{U}_{\mathrm{Q}}^H$, where $\mathbf{U}_{\mathrm{Q}}$ denotes the eigenvector matrix of $\mathbf{Q}$ and $\Lambda_{\mathrm{Q}}{=}\mathrm{diag}\left(\gamma_1, \gamma_2, \gamma_3, \gamma_4\right)$ is the diagonal eigenvalue matrix of $\mathbf{Q}$ with diagonal elements in a descending order. In the investigation that follows, based on the capacity maximization criterion, the precoding design for the $4 \times N$ LoS MIMO channels with the SWM is analyzed.

3 Precoding Design for the LoS MIMO Channels

In this section, we design the optimal precoding for the UCAs-based LoS MIMO channels with the SWM based on the maximum capacity criterion. According to the equation $\det\left(\mathbf{I} + \mathbf{AB}\right) = \det\left(\mathbf{I} + \mathbf{BA}\right)$, (2) can be equivalent to

$$C = \max_{\mathrm{Tr}(\mathbf{Q})=P} E\left[\log_2 \det\left(\mathbf{I}_2 + \frac{\mathbf{U}_{\mathrm{Q}}\Lambda_{\mathrm{Q}}\mathbf{U}_{\mathrm{Q}}^H\mathbf{H}^H\mathbf{H}}{\sigma^2}\right)\right]. \tag{3}$$

Assuming that the differences of path loss are ignored, the elements of the normalized LoS channel matrix with the SWM can be expressed as [4]

$$\left(\mathbf{H}\right)_{l,k}{=}e^{-\frac{j2\pi d_{l,k}}{\lambda}}, \tag{4}$$

where $\left(\mathbf{H}\right)_{l,k}$ is the channel coefficient between the kth $(k = 1, 2, 3, 4)$ transmit antenna and the lth $(l = 1, 2, \cdots N)$ receive antenna, $d_{l,k}$ is the distance between them and λ is the wavelength. According to the analytical methods in [4], the distance $d_{l,k}$ can be expressed approximately as

$$d_{l,k}{=}D - \frac{\left(k-1\right)d_{\mathrm{t}}R\cos\left(\frac{2\pi(l-1)}{N} - \theta_t\right)}{D}. \tag{5}$$

From (4) and (5), the matrix $\mathbf{H}^H\mathbf{H}$ is expressed as

$$\mathbf{H}^H\mathbf{H}= \begin{bmatrix} N & \sum_{i=0}^{N-1} e^{-jT\cos\left(\frac{2\pi i}{N}-\theta_t\right)} & \sum_{i=0}^{N-1} e^{-j2T\cos\left(\frac{2\pi i}{N}-\theta_t\right)} & \sum_{i=0}^{N-1} e^{-j3T\cos\left(\frac{2\pi i}{N}-\theta_t\right)} \\ \sum_{i=0}^{N-1} e^{jT\cos\left(\frac{2\pi i}{N}-\theta_t\right)} & N & \sum_{i=0}^{N-1} e^{-jT\cos\left(\frac{2\pi i}{N}-\theta_t\right)} & \sum_{i=0}^{N-1} e^{-j2T\cos\left(\frac{2\pi i}{N}-\theta_t\right)} \\ \sum_{i=0}^{N-1} e^{j2T\cos\left(\frac{2\pi i}{N}-\theta_t\right)} & \sum_{i=0}^{N-1} e^{jT\cos\left(\frac{2\pi i}{N}-\theta_t\right)} & N & \sum_{i=0}^{N-1} e^{-jT\cos\left(\frac{2\pi i}{N}-\theta_t\right)} \\ \sum_{i=0}^{N-1} e^{j3T\cos\left(\frac{2\pi i}{N}-\theta_t\right)} & \sum_{i=0}^{N-1} e^{j2T\cos\left(\frac{2\pi i}{N}-\theta_t\right)} & \sum_{i=0}^{N-1} e^{jT\cos\left(\frac{2\pi i}{N}-\theta_t\right)} & N \end{bmatrix}, \tag{6}$$

where $T = 2\pi d_t R/\lambda D$. Using the Euler formula, the entry of $\mathbf{H}^H\mathbf{H}$ can be rewritten as

$$g_{n,m}=\sum_{i=0}^{N-1}\cos\left((n-m)T\cos\left(\frac{2\pi i}{N}-\theta_t\right)\right)+j\sum_{i=0}^{N-1}\sin\left((n-m)T\cos\left(\frac{2\pi i}{N}-\theta_t\right)\right), \tag{7}$$

where $g_{n,m}$ $(n, m = 1, 2, 3, 4)$ denotes the element on the nth row and mth column of the matrix $\mathbf{H}^H\mathbf{H}$. When the number of antennas N is sufficiently large, (7) can be expressed as

$$\begin{aligned} g_{n,m} =& \frac{N}{2\pi}\int_0^{2\pi}\cos\left((n-m)T\cos(x-\theta_t)\right)dx \\ &+\frac{jN}{2\pi}\int_0^{2\pi}\sin\left((n-m)T\cos(x-\theta_t)\right)dx \\ =&\frac{N}{\pi}\int_0^{\pi}\cos\left((n-m)T\cos(x-\theta_t)\right)dx \end{aligned} \tag{8}$$

We use the method of substitution, i.e., $x-\theta_t = y-\frac{\pi}{2}$, Eq. (7) is rewritten as

$$\begin{aligned} g_{n,m} =& \frac{N}{\pi}\int_{\frac{\pi}{2}-\theta_t}^{\frac{3\pi}{2}-\theta_t}\cos\left(|n-m|T\cos(y-\frac{\pi}{2})\right)dy \\ &+\frac{N}{\pi}\int_{\frac{\pi}{2}-\theta_t}^{0}\cos\left(|n-m|T\sin y\right)dy+\frac{N}{\pi}\int_0^{\pi}\cos\left(|n-m|T\sin y\right)dy \\ =& NJ_0\left(|n-m|T\right) \end{aligned} \tag{9}$$

where $J_0(x) = \left(1/\pi\right)\int_0^{\pi}\cos(x\sin\alpha)d\alpha$ is the zero-order Bessel function of the first kind [9].

3.1 Optimal Precoding Matrix

In this section, based on the maximum capacity criterion and the entry of the matrix in (8), the optimal precoding matrix for the LoS MIMO channels with the SWM is derived.

Theorem 1 *When the parameter $T \to \infty$, i.e., $J_0(T) = J_0(2T) = J_0(3T) \approx 0$, the optimal precoding matrix becomes any orthogonal matrix. When the parameter T does not tend to infinity, the optimal precoding matrix $\mathbf{U}_{\mathrm{Q}}^{\mathrm{opt}}$ which maximizes the ergodic capacity for the LoS MIMO channels with the SWM is expressed as*

$$\mathbf{U}_{\mathrm{Q}}^{\mathrm{opt}}=\frac{1}{2}\begin{bmatrix} \frac{d}{\sqrt{c^2+d^2+\sqrt{c^2+d^2}c}} & \frac{e}{\sqrt{c^2+e^2-\sqrt{c^2+e^2}c}} & \frac{d}{\sqrt{c^2+d^2-\sqrt{c^2+d^2}c}} & \frac{e}{\sqrt{c^2+e^2+\sqrt{c^2+e^2}c}} \\ \frac{c+\sqrt{c^2+d^2}}{\sqrt{c^2+d^2+\sqrt{c^2+d^2}c}} & \frac{c-\sqrt{c^2+e^2}}{\sqrt{c^2+e^2-\sqrt{c^2+e^2}c}} & \frac{c-\sqrt{c^2+d^2}}{\sqrt{c^2+d^2-\sqrt{c^2+d^2}c}} & \frac{\sqrt{c^2+e^2}+c}{\sqrt{c^2+e^2+\sqrt{c^2+e^2}c}} \\ \frac{c+\sqrt{c^2+d^2}}{\sqrt{c^2+d^2+\sqrt{c^2+d^2}c}} & \frac{\sqrt{c^2+e^2}-c}{\sqrt{c^2+e^2-\sqrt{c^2+e^2}c}} & \frac{c-\sqrt{c^2+d^2}}{\sqrt{c^2+d^2-\sqrt{c^2+d^2}c}} & \frac{-\sqrt{c^2+e^2}-c}{\sqrt{c^2+e^2+\sqrt{c^2+e^2}c}} \\ \frac{d}{\sqrt{c^2+d^2+\sqrt{c^2+d^2}c}} & \frac{-e}{\sqrt{c^2+e^2-\sqrt{c^2+e^2}c}} & \frac{d}{\sqrt{c^2+d^2-\sqrt{c^2+d^2}c}} & \frac{-e}{\sqrt{c^2+e^2+\sqrt{c^2+e^2}c}} \end{bmatrix}, \tag{10}$$

where

$$\begin{aligned} c&=2J_0(T) - 2J_0(3T) \\ d&=4J_0(T) + 4J_0(2T) \, . \\ e&=4J_0(T) - 4J_0(2T) \end{aligned} \tag{11}$$

Proof According to the conclusion of [10], when the precoding matrix $\mathbf{U}_{\mathrm{Q}}$ is equal to the eigenvectors of the matrix $\mathbf{H}^H\mathbf{H}$, i.e., $\mathbf{U}_{\mathrm{Q}}^{\mathrm{opt}} = \mathbf{U}$, the ergodic capacity in (3) is maximized. Therefore, *Theorem* 1 can be directly derived by using the eigenvalue decomposition $\mathbf{H}^H\mathbf{H}=\mathbf{U}\Lambda\mathbf{U}^H$, where $\Lambda=\mathrm{diag}(\lambda_1, \lambda_2, \lambda_3, \lambda_4)$ is the eigenvalue matrix and $\mathbf{U}$ is the eigenvector matrix.

3.2 Optimal Power Allocation

Based on (9), the optimal power allocation for the LoS MIMO channels with the SWM is provided in the following theorem.

Theorem 2 *In the LoS MIMO channels with the proposed precoding matrix, the four cases of the analytical expressions for the optimal power allocation with the SWM are derived as follows.*

1) $\rho + \frac{1}{\lambda_1} + \frac{1}{\lambda_2} + \frac{1}{\lambda_3} \geq \frac{3}{\lambda_4}$

In this scenario, the analytical expressions for the optimal power allocation with the SWM are

$$\begin{cases} \gamma_1^{\mathrm{opt}}=\frac{P}{4}-\frac{4\sigma^2\left(2+J_0(T)+J_0(3T)-\sqrt{c^2+d^2}\right)}{16N(1+J_0(T))(1+J_0(3T))-Nd^2}+\frac{4\sigma^2(2-J_0(T)-J_0(3T))}{16N(1-J_0(T))(1-J_0(3T))-Ne^2} \\ \gamma_2^{\mathrm{opt}}=\frac{P}{4}+\frac{4\sigma^2(2+J_0(T)+J_0(3T))}{16N(J_0(T)+1)(J_0(3T)+1)-Nd^2}-\frac{4\sigma^2\left(2-J_0(T)-J_0(3T)-\sqrt{c^2+e^2}\right)}{16N(1-J_0(T))(1-J_0(3T))-Ne^2} \\ \gamma_3^{\mathrm{opt}}=\frac{P}{4}-\frac{4\sigma^2\left(2+J_0(T)+J_0(3T)+\sqrt{c^2+d^2}\right)}{16N(1+J_0(T))(1+J_0(3T))-Nd^2}+\frac{4\sigma^2(2-J_0(T)-J_0(3T))}{16N(1-J_0(T))(1-J_0(3T))-Ne^2} \\ \gamma_4^{\mathrm{opt}}=\frac{P}{4}+\frac{4\sigma^2(2+J_0(T)+J_0(3T))}{16N(J_0(T)+1)(J_0(3T)+1)-Nd^2}-\frac{4\sigma^2\left(2-J_0(T)-J_0(3T)+\sqrt{c^2+e^2}\right)}{16N(1-J_0(T))(1-J_0(3T))-Ne^2} \end{cases}. \tag{12}$$

2) $\frac{3}{\lambda_3} \le \rho + \frac{1}{\lambda_1} + \frac{1}{\lambda_2} + \frac{1}{\lambda_3} < \frac{3}{\lambda_4}$

In this case, the optimal power allocation for the LoS MIMO channels with the SWM can be expressed as

$$\begin{cases} \gamma_1^{\mathrm{opt}}=\frac{P}{3}-\frac{8\sigma^2\left(4+2J_0(T)+2J_0(3T)-3\sqrt{c^2+d^2}\right)}{96N(J_0(T)+1)(J_0(3T)+1)-6Nd^2}+\frac{4\sigma^2}{3N\left(4-2J_0(T)-2J_0(3T)+\sqrt{c^2+e^2}\right)} \\ \gamma_2^{\mathrm{opt}}=\frac{P}{3}+\frac{16\sigma^2(2+J_0(T)+J_0(3T))}{48N(J_0(T)+1)(J_0(3T)+1)-3Nd^2}-\frac{8\sigma^2}{3N\left(4-2J_0(T)-2J_0(3T)+\sqrt{c^2+e^2}\right)} \\ \gamma_3^{\mathrm{opt}}=\frac{P}{3}-\frac{8\sigma^2\left(4+2J_0(T)+2J_0(3T)+3\sqrt{c^2+d^2}\right)}{96N(J_0(T)+1)(J_0(3T)+1)-6Nd^2}+\frac{4\sigma^2}{3N\left(4-2J_0(T)-2J_0(3T)+\sqrt{c^2+e^2}\right)} \\ \gamma_4^{\mathrm{opt}}=0 \end{cases}. \tag{13}$$

3) $\frac{2}{\lambda_2} \le \rho + \frac{\lambda_1+\lambda_2}{\lambda_1\lambda_2} < \frac{2}{\lambda_3}$

For this case, the analytical expressions for the optimal power allocation are

$$\begin{cases} \gamma_1^{\mathrm{opt}}=\frac{P}{2}+\frac{2\sigma^2\left(4J_0(T)+4J_0(3T)+\sqrt{c^2+d^2}-\sqrt{c^2+e^2}\right)}{N\left(4-2J_0(T)-2J_0(3T)+\sqrt{c^2+e^2}\right)\left(4+2J_0(T)+2J_0(3T)+\sqrt{c^2+d^2}\right)} \\ \gamma_2^{\mathrm{opt}}=\frac{P}{2}-\frac{2\sigma^2\left(4J_0(T)+4J_0(3T)+\sqrt{c^2+d^2}-\sqrt{c^2+e^2}\right)}{N\left(4-2J_0(T)-2J_0(3T)+\sqrt{c^2+e^2}\right)\left(4+2J_0(T)+2J_0(3T)+\sqrt{c^2+d^2}\right)} \\ \gamma_3^{\mathrm{opt}}=\gamma_4^{\mathrm{opt}}=0 \end{cases}. \tag{14}$$

4) $\frac{1}{\lambda_1} \le \rho + \frac{1}{\lambda_1} < \frac{1}{\lambda_2}$

In this scenario, the optimal power allocation with the SWM can be expressed as

$$\begin{cases} \gamma_1^{\mathrm{opt}}=P \\ \gamma_2^{\mathrm{opt}}=\gamma_3^{\mathrm{opt}}=\gamma_4^{\mathrm{opt}}=0 \end{cases}. \tag{15}$$

Proof To prove the theorem, we firstly derive the eigenvalues by using the eigenvalue decomposition $\mathbf{H}^H\mathbf{H}=\mathbf{U}\Lambda\mathbf{U}^H$, which are expressed as

$$\begin{cases} \lambda_1 = N + \frac{NJ_0(T) + NJ_0(3T) + N\sqrt{(J_0(T) - J_0(3T))^2 + 4(J_0(T) + J_0(2T))^2}}{2} \\ \lambda_2 = N - \frac{NJ_0(T) + NJ_0(3T) - N\sqrt{(J_0(T) - J_0(3T))^2 + 4(J_0(T) - J_0(2T))^2}}{2} \\ \lambda_3 = N + \frac{NJ_0(T) + NJ_0(3T) - N\sqrt{(J_0(T) - J_0(3T))^2 + 4(J_0(T) + J_0(2T))^2}}{2} \\ \lambda_4 = N - \frac{NJ_0(T) + NJ_0(3T) + N\sqrt{(J_0(T) - J_0(3T))^2 + 4(J_0(T) - J_0(2T))^2}}{2} \end{cases}, \quad (16)$$

It is observed from (16) that when the parameter $T \to \infty$, i.e., $J_0(T) = J_0(2T) = J_0(3T) \approx 0$, the matrix $\mathbf{H}^H\mathbf{H}$ has four identical eigenvalues. Based on the eigenvalues and *Theorem* 1, the ergodic capacity optimization problem with the SWM can be rewritten as

$$\begin{array}{c} C = \max\limits_{\{\gamma_i\}_{i=1}^4} \sum\limits_{i=1}^4 \log_2\left(1 + \frac{\gamma_i\lambda_i}{\sigma^2}\right) \\ \text{s.t.}\gamma_i \geq 0, i = 1, 2, 3, 4 \text{ and } \sum_{i=1}^4 \gamma_i = P \end{array}. \quad (17)$$

The four cases for the corresponding power allocation with the SWM in *Theorem* 2 is derived by solving the optimal capacity problem in (17) with the lagrange multiplier method.

As can be seen from the expressions (12)–(15), the optimal power allocation is completely determined by $J_0(|n-m|T)$, the transmit power, the noise power and the number of receive antennas. In the fourth case, the optimal precoding becomes the beamforming algorithm along the leading eigenvector of the matrix.

4 Simulation Results

This section verifies the correctness for the theoretical analysis results and presents simulation results for the 4×100 LoS MIMO channels. The performance of the proposed precoding design and equal power allocation scheme is compared. Figure 2 shows the ergodic capacity comparison between the precoding scheme and the equal power allocation scheme for SNR in the LoS MIMO channels with the SWM, where the carrier frequency is 15 GHz. It can be seen clearly that the capacity of the proposed precoding scheme is better than that of the equal power allocation scheme. With the decrease of the parameter T, the difference between the ergodic capacity of the proposed scheme and the equal power allocation scheme is obviously increased. When the parameter $T \to \infty$, i.e., $J_0(T) = J_0(2T) = J_0(3T) \approx 0$, the matrix $\mathbf{H}^H\mathbf{H}$ becomes the matrix $N\mathbf{I}_4$ and has four identical eigenvalues. Then the capacity of the proposed precoding scheme is equal to that of the equal power allocation scheme.

Figure 3 shows the relationship between the capacity and the parameter T in the 4×100 LoS MIMO channels. When SNR is 5 dB or 10 dB, it can be seen from Fig. 3 that the capacity increases with the increasing parameter T and tends to be

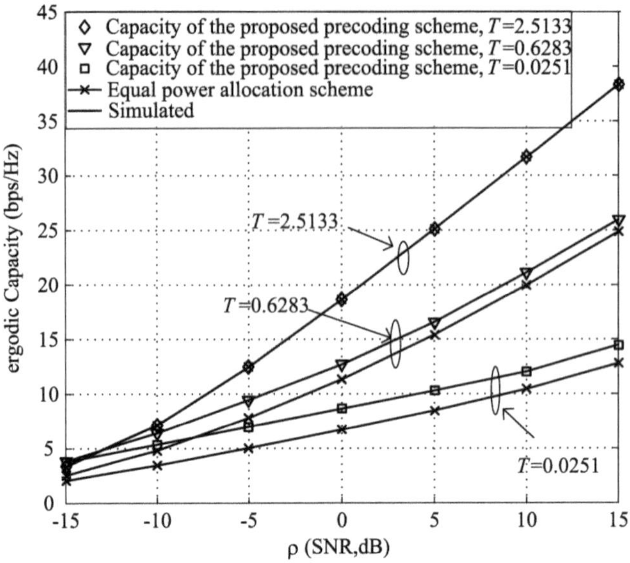

Fig. 2 Comparison of the LoS MIMO channel capacity for the proposed precoding design and equal power allocation scheme

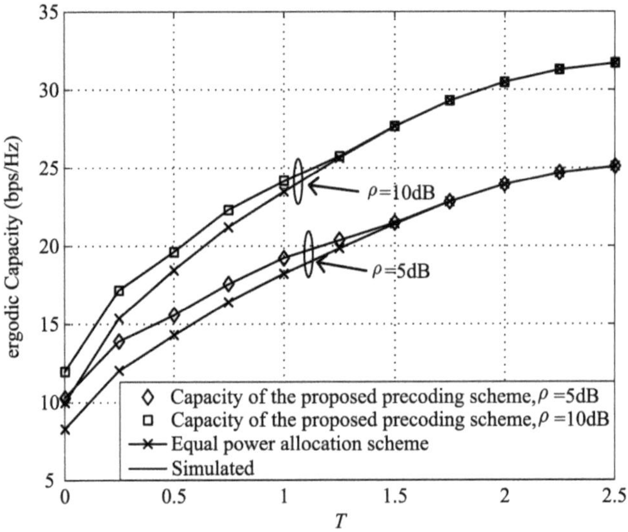

Fig. 3 Relationship between the LoS channel capacity and parameter T, where $\rho = 5\text{dB}$ and $\rho = 10\,\text{dB}$

stable. In MIMO channels, the degree of freedom is a key performance measure for the channel capacity. When the parameter $T \rightarrow \infty$, the $4 \times N$ LoS MIMO channels have four spatial degree of freedom, and the corresponding capacity tends to a saturation value.

5 Conclusion

The precoding for the $4 \times N$ LoS MIMO channels with the ULAs at the Tx and UCAs at the Rx is studied in this paper. By maximizing the ergodic capacity, the closed-form expressions for the precoding matrix and power allocation in LoS MIMO channels with the SWM are derived. Simulation results show that the proposed precoding scheme can obtain better performance compared with equal power allocation scheme. In addition, the transceiver distance, wavelength, array element spacing and the number of receive antennas play a decisive role in the proposed scheme.

Acknowledgements This work was supported by the National Natural Science Foundation of China (61701062) and the Doctoral Startup Found of Chongqing University of Posts and Telecommunications (A2016-15).

References

1. Jiang, J.S., Ingram, M.A.: Spherical-wave model for short-range MIMO. IEEE Trans. Commun. **53**(9), 1534–1541 (2005)
2. Sarris, I., Nix, A.R.: Design and performance assessment of high-capacity MIMO architectures in the presence of a line-of-sight component. IEEE Trans. Veh. Technol. **56**(4), 2194–2202 (2007)
3. Skentos, N., Kanatas, A.G., Pantos, G.: Capacity results from short range fixed MIMO measurements at 5.2 GHz in urban propagation environment. In: IEEE International Conference on Communications, pp. 3020–3024. IEEE, Paris (2004)
4. Pu, X., Shao, S., Deng, K.: Analysis of the Capacity Statistics for 2 2 3D MIMO Channels in Short-Range Communications. IEEE Commun. Lett. **19**(2), 219–222 (2015)
5. Tamaddondar, M.M., Noori, N.: Plane wave against spherical wave assumption for non-uniform linear massive MIMO array structures in LOS condition. In: Iranian Con-ference on Electrical Engineering (ICEE), pp. 1802–1805. IEEE, Tehran (2017)
6. Pu, X., Chen, Q., Shao, S.: Transmit design for short-range MIMO channels with the spherical-wave model. IEEE Commun. Lett. **21**(8), 1875–1878 (2017)
7. Constantine, A.B.: Antenna Theory: Analysis and Design, 2nd edn. Wiley, Hoboken (1997)
8. Paulraj, A., Nabar, R., Gore, D.: Introduction to Space-Time Wireless Communications. Cambridge University Press, Cambridge (2003)
9. Temme, N.M., Zwillinger, D.: Special Functions: An Introduction to the Classical Functions of Mathematical Physics, 2nd edn. American Journal of Physics (1997).
10. Telatar, E.: Capacity of multi-antenna Gaussian channels. Eur. Trans. Telecommun. **10**(6), 585–595 (1999)

An Envisioned Virtual Gateway Architecture for Capillary Networks in Smart Cities

Deze Zeng and Lin Gu

1 Introduction

"Better City, Better Life" has been becoming an ever-pursuing vision of human beings. The rosy prospect has attracted much attention from different sectors. According to a recent smart city report from Pike Research, it is estimated that around 16 billion dollars will be invested annually on smart city market by 2020 and the total budget will go for hundreds of billions. Many smart city projects, e.g., European smart cities project,[1] have already taken off.

Information and communication technology (ICT) plays an important role in improving the cities' productivity, efficiency, sustainability and quality-of-life. Especially, with the popularity and fast development of Internet-of-Things (IoT), which is an infrastructure interconnecting a large number of smart devices (e.g., sensors) and is able to collect massive data from the physical-world. It is predicted that over 20 billion devices will be connected in IoT by 2020. Hence, a more aggressive vision is proposed as Internet-of-Everything (IoE) [6]. IoT is the key enabling technology to the smart city vision. By deploying urban IoT, we obtain an infrastructure that is able to integrate the physical-world into the cyber-world, inspiring smart city application innovations. With the convergence of physical-world and

[1] http://www.smart-cities.eu.

D. Zeng (✉)
School of Computer Science, China University of Geosciences, Wuhan, China
e-mail: deze@cug.edu.cn

L. Gu
School of Computer Science and Technology, Huazhong University of Science and Technology, Wuhan, China

B. Ye et al. (eds.), *2nd International Conference on 5G for Ubiquitous Connectivity*, EAI/Springer Innovations in Communication and Computing,
https://doi.org/10.1007/978-3-030-22316-8_12

cyber-world, a variety of smart city applications in different domains, e.g., structural health of buildings, waste management, noise monitoring, traffic management, city energy consumption, etc., can be developed, as summarized in [15].

In IoT, Machine-to-Machine (M2M) communication, a.k.a., machine type communication (MTC), is indispensable. A diversity of radio access technologies (RATs) or standard protocols, e.g., Wi-Fi HaLow, LTE-MTC, Bluetooth Low Energy, ZigBee, GPRS, LoRa, SigFox, NB-IoT, etc., have been proposed. Compared with human-to-human or human-to-machine communications, the prominent features of MTC are low-power and low-rate as smart devices are usually battery powered but require long operation time in terms of years.

Network consisting of smart devices interconnected via short-range communications is termed as *capillary network* in the literature, comparing with the tiny blood vessels connecting cells to the arteries [1, 3, 9, 11]. Augé-Blum et al. [1] identify that capillary network caters to the emerging needs of smart cities as it unifies the wealth of wireless connectivity available in urban environment. Sachs et al. [11] think that capillary networks and short-range communications are key development in the networked society as they provide connectivity for billions of devices in many use cases, e.g., electricity meter monitoring. Novo et al. [9] believe that capillary networks bridge cellular and IoT worlds, and therefore will be a fundamental part of the loT development.

Resource-limited smart devices cannot bear full Ethernet protocol stack and usually are not IP-based. While, smart city applications strongly require a ubiquitous connection of all devices. To address these issues, gateways are introduced. A gateway is a specialized device locating between capillary network and the Internet to interpret the protocols for the two sides. However, with the increasing needs from smart city applications, existing capillary networks suffer following limitations.

- *Interoperatability*: Different vendors deploy capillary networks with different RATs. The coexistence of manifold networking interfaces imposes obstacle for the interoperability or cooperation between capillary networks, hindering smart city application innovation.
- *Maintenance*: In a large-scale capillary network, it is inevitable that some components may unexpectedly fail due to certain reasons. Besides, we may need to periodically upgrade the smart city systems by adding in or upgrading some components to catch up with the increasing demands or adapt to the changing surrounding environment. At this time, changing or upgrading the hardware component could be a tedious task.
- *Scalability*: Limited by the service capacity of a gateway, it cannot simultaneously serve unlimited number of devices or networks. Whenever we need to augment a new system to our smart city, we need to specially deploy a network as well as the associated gateway. Such one-to-one mapping limits the scalability of the smart city network.
- *Flexibility*: Capillary network gateway must be aware of, and loyal to, the RAT used by the networked devices as well as its prosperities. When another new capillary network with different RAT is added in, existing gateway usually

is not able to support the newly added one. We may have to upgrade the gateway or deploy a new one. This is especially frustrating in smart city scenario where a diversity of RATs are used in parallel, and new or customized RAT is continuously introduced and developed. Besides RAT, the varying demands in capillary networks also require the gateway to adapt to the service needs. Whenever a new service is required, certain functionalities, e.g., in-network processing, shall be enabled at the gateways. Hence, the gateway shall be flexible enough for the developers to customize or deploy functionalities on it. Besides radio access, the varying demands in capillary networks also requite the gateway to be able to adapt to the service needs. Whenever a new service is required, certain functionalities, e.g., in-network processing, shall be enabled at the gateways. The gateway shall be flexible enough for the developers to customize or deploy functionalities on it.
- *Mobility*: In some urban smart city applications (e.g., participatory sensing), the communication nodes may move randomly. This requires that the gateway shall intelligently move with the mobile nodes or the mobile nodes can dynamically switch between gateways, to ensure that the data can be efficiently uploaded to the cloud.
- *Expenditure*: Deploying, maintaining and upgrading large-scale capillary network is not only tedious but also with high cost, especially for vendor-specific hardware based implementation. For a long-term smart city application, the operational expenditure may even exceed the capital expenditure.

Given the above limitations of the last-mile capillary networks for smart cities, proposing new open, scalable, flexible and cost-efficient architecture is significant. In this article, motivated by the newly emerging technology such as software-defined networking (SDN), network function virtualizaiton (NFV) and cloud radio access network (CRAN), we propose a virtual gateway architecture for capillary networks, with the principle of detaching and softwarizing certain functions from traditional hardware gateways. The remainder of this paper is organized as follows. We first depict the architecture of capillary networks and specially introduce the role of gateways in capillary networks. The limitations of existing gateway implementation methods are then outlined. To tackle these limitations, we propose a virtual gateway design for capillary networks and discuss its applicability and implementation feasibility.

2 Capillary Network

2.1 Architecture

Capillary networks intend to interconnect various tiny devices and provide strong support to realize a digital networked society where citizens can obtain information from these devices or even actuate them according to application needs. A typical urban capillary network consists of a set of smart tiny devices with different capabilities. For example, a capillary network for gas leakage detection and prevention in a building may include devices such as gas usage monitors, gas leakage detectors, gas sensor and automotive valve.

These tiny devices are usually interconnected using short-range RATs. Besides standard RATs, developers may also customize their own RAT according to the application needs. Nevertheless, different RATs are with different specifications or mechanisms on medium access, routing, scheduling, error detection and correction, etc. This raises network-dimensioning challenges to smart city application innovation. From the engineering perspective, application developers or system administrators need to take inter-technology inter-operation into consideration.

To access the Internet, these devices shall be connected to a more powerful device called capillary gateway, which further connects to the Internet via fronthaul networks (e.g., Ethernet, optical network, cellular network). The gateway is able to process and forwards data to the cloud. The cloud hosting various smart city applications that digest the data from capillary networks and generate response accordingly. The response (e.g., actuation command) is then transported back to the capillary network via the gateway. Note that there is an emerging trend in the development of capillary networks is to eliminate the use of gateways by enabling IPv6 directly on the smart things. For example, the Internet Engineering Task Force (IETF) has prepared proposed new protocols, e.g., 6LoWPAN, describing how to transport IPv6 over IEEE 802.15.4 and BLE. However, as argued by Zachariah et al. in [14], simply enabling IPv6 connectivity alone falls short of the full set of possibilities of a true IoT gateway. Therefore, we may still need gateways for capillary networks in some cases.

Figure 1 illustrates a smart city scenario supported by three capillary networks, which adopt MTC RATs Bluetooth, ZigBee and LTE-MTC, respectively. Each

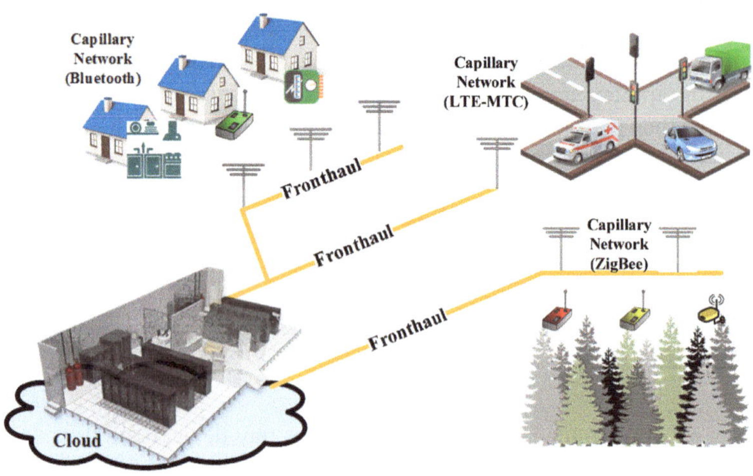

Fig. 1 Capillary networks in smart city

network has one gateway connected to the cloud via Ethernet, optical and LTE, respectively. Gateway is a key component of capillary network as it effectively migrates the variety and diversity of the devices and makes a bridge between the smart devices and the Internet.

2.2 Capillary Network Features

According to the above description to capillary networks in smart cities, we can see that capillary networks exhibit the following special features.

- *RAT heterogeneity*: As we have known, there are already many different commercially recognized RATs to support various capillary networks with different needs. It is worth noting that there will be more new RATs to be developed to catch up with the increasing demands from smart city applications. Such heterogeneity creates a chaotic environment that shall be handled by capillary network gateways.
- High density: High-density is one of the most prominent features of capillary networks as the network for a specific application may consists of a large number of MTC devices.
- Low rate: Compared with human-centric communications, MTC devices (e.g., smart meter) are usually with low data rate. For example, structural health monitoring in Padova Smart City project[2] requires only 1 packet every 10 min per device [15].
- Low power: Most MTC devices are battery-powered. Although some recent proposals advocated the adoption of renewable energy harvesting technique, the available energy in a device is still limited by the harvesting ability and instability of energy source.
- QoS diversity: Compared with human-centric communications, the QoS requirements of capillary networks are more diverse. That is, the QoS requirements vary tremendously from one network to another, depending on the applications. Some applications may require guaranteed network connectivity while some others may ask for extremely low latency. For example, a sensor for fire prevention should be able to reliably transmit data in a short time after the detecting an anomaly.
- Multihop: Capillary networks can organize in an ad hoc manner. Addressing and routing are two essential issues. Addressing in the capillarity, e.g., in 6LoWPAN or ZigBee, is different to the one used on Internet. Therefore, a gateway with the capability of address conversion is needed. Besides, in order to minimize the expense for data querying in a multihop capillary network, the gateway can also cache previously visited data to minimize the request of data to the sensors.

[2]https://eu-smartcities.eu/place/padova.

3 Existing Gateway Implementation Methods

In this section, we briefly review some existing gateway implementation methods and standards.

Today most of the gateways are based on a program that runs on a customized firmware. An embedded firmware is usually designed and implemented by gateway manufacturer. Essential gateway functions such as wireless communication, network control and management are usually embedded in the gateway as ASIC. Limited customization ability, depending on the openness of the gateway, is provided by allowing network administrators to alter some configurations through provided portal.

The embedded firmware is usually managed by an operation system. For complex applications like car infotainment, full operation systems like Linux and Windows Embedded are used. Besides, smaller footprint systems like Intel Galileo, VxWorks, QNX are often used for specialized applications such as safety, realtime monitoring. Linux is the most widely used open-source operation system. Many embedded systems are designed based on it, and gateway is no exception. Lots of efforts have already been made in designing Linux based gateway platforms, e.g., WRT54G and OpenWRT.[3]

Besides specific vendor design gateway devices, we may also utilize general CPE (Customer-premises equipment or customer-provided equipment) devices that locates at a user's premises to implement gateways. Without doubt that the most popular and well-known CPE is smartphone, whose recent prosperousness has made it ubiquitous. With the consideration of both Internet access and its ubiquity, smartphone is regarded as an ideal gateway development platform candidate for providing connectivity anytime, anywhere, and for anyone. A large number of third party softwares can be installed in smartphones to transform them as gateways for capillary networks.

To develop gateways, several standards or specifications have been proposed and adopted. OSGi (Open Service Gateway Initiative) specifies a modular development framework for Java language and provides a good framework for Java programmers to develop and deploy service-oriented modular applications and libraries. OSGi already have many open-source realizations such as Knoflerfish,[4] Equinox[5] and Apache's Felix.[6] Based on OSGi, HGI[7] (Home Gateway Initiative) intends to

[3]https://openwrt.org/.

[4]http://www.knopflerfish.org/.

[5]http://www.eclipse.org/equinox/.

[6]http://felix.apache.org/.

[7]http://www.homegatewayinitiative.org/.

integrate the OSGi platform into home gateway so as to create an environment facilitating home gateway development. By such means, gateway applications are developed as bundles, which can be remotely and dynamically installed, started, stopped, updated, and uninstalled, without touching the underlying firmware image.

Although gateway plays an essential role in capillary networks, many problems still exist. As indicated by Zachariah et al. [14], the IoT has a gateway problem to provide application-specific connectivity to IoT devices because today's gateways conflate network connectivity, in-network processing, and user interface functions. Beijar et al. [3] think that capillary networks have gateway selection problem in the consideration of multiple gateway coexistence, load balancing, end-to-end path optimization.

4 Virtual Gateway Supported Capillary Networks

We are entering the era of the customized intelligent capillary network gateway. Taking the RAT diversity into consideration, customizable solutions with comprehensive connectivity are clearly required. In this section, by applying the recently emerged networking technologies such as SDN, NFV and CRAN, we propose a virtual gateway supported capillary network architecture. We will find out that many limitations discussed above can be well addressed by this novel architecture.

4.1 Enabling Technology Overview

Software Defined Networking (SDN) decouples the control-plane from the data plane. The control plane are moved to a logically centralized controller that can reside in a datacenter. While, the data plane remains in the transport network and manipulated by the controller. SDN enables easy network programmability for flexible network management, e.g., steering the data traffic path. For example, Hampel et al. [7] applying SDN to enable vertical forwarding so as to replace specialized gateways with virtualized controllers and commoditized forwarding elements.

NFV is proposed to softwarize existing hardware-based network functions and make them be able to operate on commodity hardware such as X86 servers in the cloud. Such function cloudification method makes the network architecture highly flexible as the network can be reconfigured quickly and adaptively. The prospect of NFV has raised many discussions in the literature. Baumgartner et al. [2] discuss a mobile core network function placement with the consideration of interconnections towards the radio access network and the Internet as well as the traffic routing between the virtualized network functions.

Specially, CRAN, as NFV at the wireless edge, is proposed to split the remote radio head (RRH) from the baseband unit (BBU). BBU for radio access processing is softwarized and can reside in a centralized data center. CRAN has also widely been discussed. Beyene et al. [4] propose a software-defined radio-based architecture and implement an indoor distribute-antenna system (DAS) by using a Cloud-RAN platform. Peng et al. [10] present H-CRAN (Heterogeneous Cloud Radio Access Networks) to deal with the heterogeneity of radio access in 5G networks.

4.2 Design Overview

By exploring the above mentioned enabling technology, we present virtual gateway supported capillary networks. Its high-level overview, as compared with traditional architecture, is illustrated in Fig. 2.

As we have known, a gateway shall first be able to communicate with the smart devices with diverse RATs, i.e., heterogeneous RAT convergence. In contrast to traditional gateway which contains completed protocol stack, our design follows CloudMAC concept [5] and decouples certain functions that can be softwarized from the gateway. For each protocol, the MAC processing is partially migrated out from the gateway to the backend, making the frontend become slim access point (AP), or an antenna. These slim APs form a DAS.

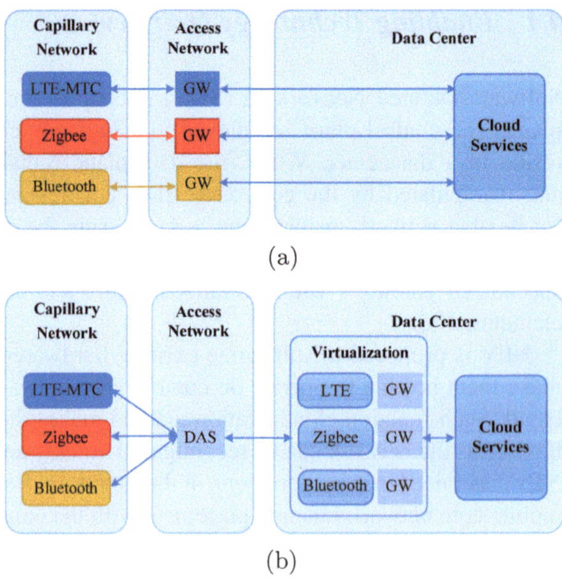

Fig. 2 Overall architecture of virtual gateway supported capillary networks, in comparison with traditional architecture. (**a**) Overview of traditional architecture. (**b**) Overview of virtual gateway supported architecture

On the uplink, the frontend simply forwards raw frames to the corresponding backend for further processing. While on the downlink, the corresponding backend generates MAC frames and beacons, adds control headers or optionally encrypts the frames, and forwards the encapsulated frames to the frontend, which then transmits those frames into capillary network with the specified transmission power, modulation/encoding schemes. The frontend performs actual RAT with the smart devices in capillary networks. A frontend antenna only receives and transmits raw MAC frames. Most MAC frames to be sent are generated by the backend server while some frames with hard-real time constraints (e.g., beacon) can be generated directly by the antenna. Therefore, we say the MAC layer is partially migrated to the cloud.

Throughout the process, we can see that no change is required on the protocol stack on the smart devices, which still use their original stack to communicate with each other, and with the gateway. How the gateway behaves is transparent to the smart devices. Within each capillary network, direct M2M communications are still in their conventional way.

Note that, a backend instance can locate anywhere, either together with the frontend or in remote cloud. Decoupling the backend does not mean making it distant from the frontend. They could also locate together and this does not influence the flexibility provided that we can access and control the backend in a softwarization way.

A slim AP may connect to one or many virtual gateways. This because different protocol software packages may reside on different virtual gateways for respective communication needs. To support a specified RAT, the frontend must connect to the virtual gateway with the corresponding protocol package. The frontend is armed with an SDN-enabled switch (e.g., OpenFlow switch), which contains a forwarding table specifying where (e.g., the backend) a frame shall be forwarded. The SDN controller runs applications that configure the forwarding table using OpenFlow protocol. By setting up the forwarding rule, the frames from one capillary network can be routed to its associated virtual gateway. As one antenna may be shared by multiple capillary networks, the entries in the table are masked by the network identifier (e.g., PAN identifier for ZigBee) and MAC address. Therefore, we can see that the binding relationship between a backend and a capillary network is determined by the forwarding entry in the table.

A unified interface to control both the frontend and backend of each capillary network is provided in the control layer. The controller is integrated with SDN controller, which is able to manage the packet manipulation behaviors (e.g., forwarding, altering, discarding, etc.) on OpenFlow enabled switches. OpenFlow switch in the backend is embedded for easy inter-network cooperation and interaction with the cloud services. The controller locates in a centralized server, e.g., in a datacenter. One controller may simultaneously manage a number of capillary networks.

To enable efficient network management and control, northbound interface is provided to network operators. Capillary network applications can be developed using these friendly high-level specifications (e.g., data and functions). Via these interfaces, virtual gateway can be created, migrated or deleted. The behaviors of a virtual gateway can be easily manipulated.

4.3 *Functions in Virtual Gateway*

We then detail the essential functions of virtual gateway, which are depicted in Fig. 3.

Frontend—Radio Access

A growing number of RATs are adopted by various intelligent devices in smart city applications. To accommodate all these protocols, current practice is to have all these protocols in a gateway and use separate antenna for each protocol. This limits the network extendability and flexibility. An ideal solution is to have a set of distributed antennas, i.e., DAS, that can be shared any protocol. This design is inspired by CRAN technique where a gateway is logically split into hardware-only radio heads and software-based components for baseband processing as well as some other network functions. By detaching and virtualzing the protocol stack, any

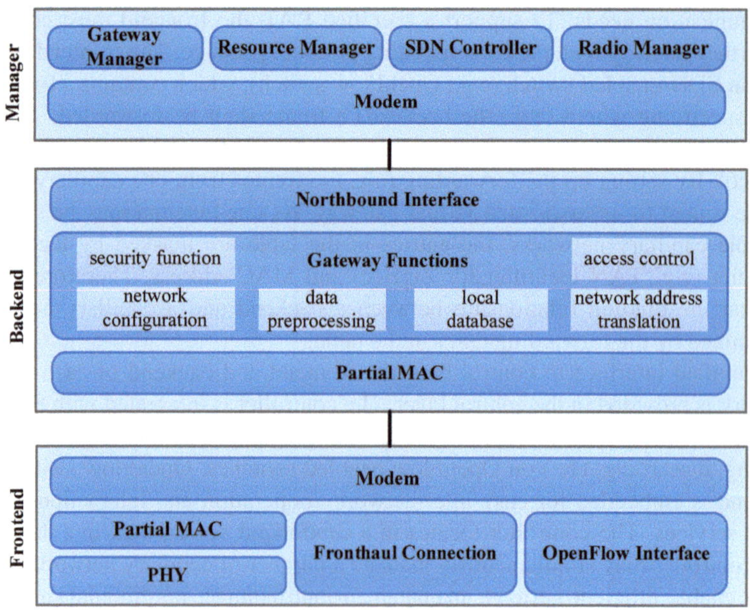

Fig. 3 Functions in virtual gateway for capillary networks

RAT can be supported provided that the corresponding softwarized protocol stack resides in the cloud. This also implies that we may even customize a new RAT by introducing a new radio interface protocol. By such means, a capillary network can adopt any customized RAT. While, this is regarded as difficult or even impractical in conventional hardware gateways.

Fronthaul Connection

Basically, a gateway acts as a bridge and is responsible for interpreting communication protocols between the managed capillary network and the Internet. Normal existing protocol stack is applied at the frontend for connection with the backend. This is similar to existing hardware based gateway architecture where a gateway can connect to the Internet with either wired or wireless connection, e.g., radio-over-fiber, Ethernet, WiFi, 2G/3G/4G, satellite, etc. For example, if optical connection is applied, layer 2 Ethernet fiber connection functions (e.g., optical network service) shall be embedded in the frontend while it shall includes an LTE interface compliant with the LTE standard for LTE-based fronthaul connection. The fronthaul connection is transparent to the RATs.

Network Functions

Besides the "bridge" function for protocol translation and address conversion, a gateway may have many embedded functions for handling local network services such as security function, network configuration, local database, data preprocessing (e.g., data filter, aggregation, etc.), access control, data caching and so on. These functions are the heart of the gateway. With the proposed virtual gateway concept, these functions can be freely defined in the cloud side according to the network management or application needs. Some essential gateway functions are shown in Fig. 3. All these functions can be softwarized and cloudified, significantly improving the easiness of network management.

4.4 *Management*

For network innovation, it is desirable that the management is transparent to capillary network operators. Therefore, some essential network management functions are introduced and shown in Fig. 3.

Radio Manager

Centralizing and pooling the radio resources imply more efficient and easier radio resource management. However, considering the coexistence of multiple RATs and the limitation of available radio resource, how to allocate the radio resources according to the application needs is still a critical issue. For example, multiple capillary networks with different RATs may share one antenna, it is essential to carefully allocate the radio resources among these capillary networks according to their needs. Radio manager allows network developers to customize the radio resource allocation scheme.

Resource Manager

AS virtual gateway may reside in a virtual machine, like common cloud services, resource management is an inevitable issue. As many functions are either computation-intensive or IO-intensive, with the consideration of resource capacity limitation, the resources shall be carefully and adaptively allocated among the gateways. Common cloud resource management framework like OpenStack or Open Nebula can be applied for resource allocation and management, e.g., VM creation and migration.

Gateway Manager

Besides normal cloud resource, capillary network management further requires management on the virtual gateways, frontend radio resource, capillary network node behaviors, to cater for the varying needs of capillary networks. To this end, gateway manager is introduced for network administrators to dynamically deploy new virtual gateways, migrate a gateway, add in new network functions, alter gateway configurations and so on.

SDN Controller

To cope with the frame or packet manipulation behaviors on the frontend and the backend, respectively, SDN controller shall be implemented. For the frontend, we shall set up forwarding rules such that the frames from a capillary network can be correctly sent to its associated backend. While, backend may also need correct forwarding rule for connection with the cloud, as well as other capillary networks for inter-operation.

4.5 Advantages and Disadvantages

Compared with traditional gateway implementations, the proposed virtual gateway architecture have the following advantages.

- *Efficiency*: The resources consumed by virtual gateways can be efficiently managed at the data center. For example, we can suspend unused virtual gateway instances and dynamically allocate resources to the gateways according to their communication needs. This results in efficient use of resources.
- *Flexibility*: Our proposal is less dependent from hardware manufacturers. New gateways can be deployed in an on-demand manner according to the application needs. In addition, gateways can be replicated or migrated by allocating required network and communication resources in the cloud.
- *Programmability*: The gateways can be reprogrammed on-the-fly according to the realtime communication demands from capillary networks. Such means promises network innovation for capillary networks.
- *Economy*: Both the capital expenditure (CAPEX) and operational expenditure (OPEX) for service providers can be significantly lowered. These is no need to deploy large number of full-fledged gateways but can explore existing antennas by implementing virtual backends. Resources allocated to the backends can be also dynamically adjusted.
- *Friendliness*: The maintenance, update and replacement and virtual gateways are similar to virtual machine management. With proper primitives, it is easy to manage the virtual gateways as how we do for virtual machines in data centers.
- *Robustness*: With the virtual gateway concept, the location of the virtual gateway is immaterial to its connectivity in the capillary networks. By reprogramming the SDN switches, data flows can be redirected to a new location. Therefore, multiple virtual gateways can be created for one capillary network. This significantly improves the system robustness as the data flow through the primary gateway can be reprogrammed to a redundant one in the failover situation.

Besides the above advantages, same as any other CRAN-based design, we have to admit that several disadvantages are also imposed.

- *Latency*: One notorious disadvantage of CRAN is the long network delay due to the comparatively slow packet processing in general purpose processor and the communication latency between the frontend and the backend server. Fortunately, pioneering researchers have proposed various means to address this issue. For example, Tandon et al. [13] consider deploy backend servers in fog computing environment, i.e., fog radio access network, to explore its distributiveness and proximity features.
- *Security*: The flexibility feature is a double-edge sword as it also raises the security concerns. With the ability to customize radio access behavior, the gateway becomes vulnerable to malicious attackers who intends to eavesdropper the others' information through the same gateway. More efforts are expected to strengthen the security in capillary gateways.

5 Implementation Feasibility and Example of Using Virtual Gateways

In this section, we first present the feasibility of realizing the proposed virtual gateway concept and then outlines a normal procedure to describe how to implement a new RAT for a capillary network in the proposed architecture.

5.1 Implementation Feasibility

CRAN, which decouples baseband processing and radio units, has already widely verified by many practical systems such as FluidNet [12] and CloudMAC [5]. In order to allow the DAS shared by coexisting heterogeneous RATs, we also expect the vision of one-antenna-multiple-protocol, which has already been demonstrated as practical by Hong et al. [8]. They design and implement Picasso that partitions fragmented spectrum into multiple slices, each of which operates concurrent and independent protocol stacks. By such means, it allows simultaneous transmission and reception on separate and arbitrary spectrum fragments using a single antenna. Our design integrates both the concept of CRAN and Picasso, enabling many-to-many mapping between the radio antennas and the RATs.

The frontend and backend can be connected via layer 2 tunnels and an OpenFlow switch. Many layer 2 tunnels protocols, e.g., Layer 2 Tunnel Protocol (L2TP), Layer 2 Forwarding Protocol (L2F) and Point-to-Point Tunneling Protocol (PPTP), can be applied OpenFlow interface can be implemented using open source Open vSwitch (OVS[8]). As part of the MAC protocol is executed at the frontend radio, MAC address can be utilized to mark a data flow. Therefore, flow differentiation can be supported via L2TP pseudo-wires by encapsulating frames with the network identifier and the MAC address. FluidNet [12] already validates the feasibility of realizing a reconfigurable mapping between the frontend and the backend.

The backend servers can run as VMs on any hypervisor, e.g., KVM, Xen, VMWare. For example, a Linux VM could be treated as a backend server for implementing virtual gateway. Besides, it has already proved that it is feasible to adopt lightweight containers (e.g., Linux Container LXC[9] and Docker[10]) with real-time kernels for fast packet processing. We may implement RAT protocol, e.g., Bluetooth, as driver in the backend server to enable the corresponding RAT at the frontends. Many open source standard protocol drivers, e.g., BlueZ,[11] are available and it is not difficult to deploy new driver for new RAT.

---

[8]http://openvswitch.org/.

[9]https://linuxcontainers.org/.

[10]http://www.docker.com/.

[11]http://www.bluez.org/.

5.2 Example of Deploying a New Capillary Network Using Virtual Gateway

Figure 4 presents the process for deploying a new capillary network with customized MTC protocol. Suppose that we are going to deploy a capillary network with a large number od distributed wireless devices for a smart city application. A customized MTC protocol is embedded in these devices. To enable interaction between the Internet and the capillary network, we need gateways that support the customized protocol. Traditionally, we shall deploy corresponding customized gateways. Now, suppose we already have several distributed antennas in the city. Let us see how we can utilize these antennas to enable intelligent virtual gateways using the proposed method.

At first, software package corresponding to the MTC protocol shall be implemented. To host the software package, we can start a new virtual machine in the cloud. By integrating with the software package, the virtual machine is able to interpret the frames following the customized MTC protocol. Besides radio access function, we shall also design and implement other gateway functions such as data preprocessing in the backend according to the capillary network application needs.

The frames sent from devices in the capillary network shall be correctly forwarded to the implemented backend. Corresponding forwarding rule shall be inserted in the forwarding table in the shared frontend. We can use the MAC address of the virtual machine as the PAN identifier for setting up the frame forwarding rule. To this end, the forwarding policy shall be enforced by SDN controller, which further sets up forwarding rule in the frontend accordingly. If we need multiple frontends to support the capillary networks, we simply program the switches in all wanted frontends to make them connect to the backend.

Once both the frontend and backend are set up, we can deploy capillary network devices and let them work on the customized MTC. The generated frames can

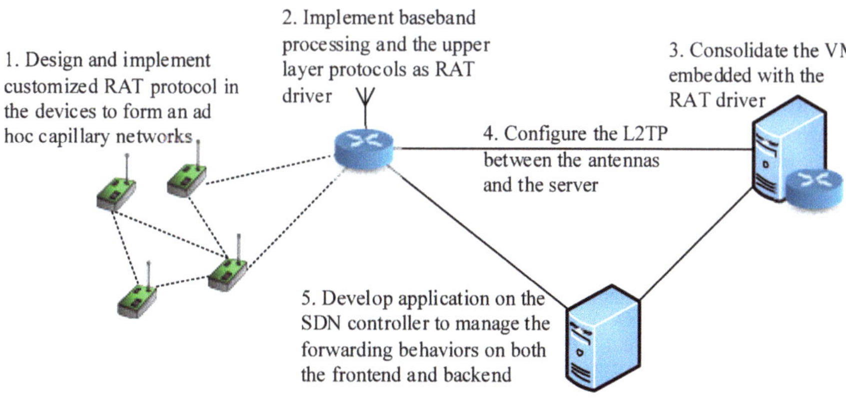

Fig. 4 Create a virtual gateway and deploy a new capillary network with customized RAT

be freely sent to any frontend connecting with the backend hosting the protocol. By now, developers can develop any wanted smart city applications using the northbound interfaces on the backend.

6 Conclusion

In this article, we propose a virtual gateway supported capillary network architecture for smart city by applying SDN, NFV and CRAN. In our proposed architecture, certain functions that can be softwarized are decoupled and migrated out. This significantly improves the network flexibility in accommodating a diversity of RATs. With the recent progress in SDN, NFV and CRAN, we believe that our proposal is feasible and ideal for future smart city application innovations.

Acknowledgment This research was supported by the NSF of China (Grant No. 61772480, 61602199, 61672474, 61402425).

References

1. Augé-Blum, I., Boussetta, K., Rivano, H., Stanica, R., Valois, F.: Capillary networks: a novel networking paradigm for urban environments. In: Proceedings of the 1st Workshop on Urban Networking, pp. 25–30. ACM, New York (2012)
2. Baumgartner, A., Reddy, V.S., Bauschert, T.: Mobile core network virtualization: a model for combined virtual core network function placement and topology optimization. In: Proceedings of the 1st IEEE Conference on Network Softwarization (NetSoft), pp. 1–9. IEEE, Piscataway (2015)
3. Beijar, N., Novo, O., Jimenez, J., Melen, J.: Gateway selection in capillary networks. In: Proceedings of the 5th International Conference on the Internet of Things, pp. 90–97. IEEE: Piscataway (2015)
4. Beyene, Y.D., Jantti, R., Ruttik, K.: Cloud-RAN architecture for indoor DAS. IEEE Access **2**, 1205–1212 (2014)
5. Dely, P., Vestin, J., Kassler, A., Bayer, N., Einsiedler, H., Peylo, C.: CloudMAC - an openflow based architecture for 802.11 MAC layer processing in the cloud. In: Globecom Workshops (GC Wkshps), pp. 186–191. IEEE, Piscataway (2012)
6. Evans, D.: The internet of everything: how more relevant and valuable connections will change the world. In: Cisco IBSG, pp. 1–9 (2012)
7. Hampel, G., Steiner, M., Bu, T.: Applying software-defined networking to the telecom domain. In: IEEE Conference on Computer Communications Workshops (INFOCOM WKSHPS), pp. 133–138. IEEE, Piscataway (2013)
8. Hong, S.S., Mehlman, J., Katti, S.: Picasso: flexible RF and spectrum slicing. ACM SIGCOMM Comput Commun Rev **42**(4), 37–48 (2012)
9. Novo, O., Beijar, N., Ocak, M., Kjallman, J., Komu, M., Kauppinen, T.: Capillary networks-bridging the cellular and IoT worlds. In: Proceedings of the 2nd World Forum on Internet of Things, pp. 571–578. IEEE, Piscataway (2015)
10. Peng, M., Li, Y., Zhao, Z., Wang, C.: System architecture and key technologies for 5G heterogeneous cloud radio access networks. IEEE Netw **29**(2), 6–14 (2015)

11. Sachs, J., Beijar, N., Elmdahl, P., Melen, J., Militano, F., Salmela, P.: Capillary networks–a smart way to get things connected. Ericsson Rev **91**, 12–19 (2014)
12. Sundaresan, K., Arslan, M.Y., Singh, S., Rangarajan, S., Krishnamurthy, S.V.: FluidNet: a flexible cloud-based radio access network for small cells. In: Proceedings of the 19th Annual International Conference on Mobile Computing & Networking, pp. 99–110. ACM, New York (2013)
13. Tandon, R., Simeone, O.: Cloud-aided wireless networks with edge caching: fundamental latency trade-offs in fog radio access networks. In: 2016 IEEE International Symposium on Information Theory (ISIT), pp. 2029–2033 (July 2016)
14. Zachariah, T., Klugman, N., Campbell, B., Adkins, J., Jackson, N., Dutta, P.: The internet of things has a gateway problem. In: Proceedings of the 16th International Workshop on Mobile Computing Systems and Applications, pp. 27–32. ACM, New York (2015)
15. Zanella, A., Bui, N., Castellani, A., Vangelista, L., Zorzi, M.: Internet of things for smart cities. IEEE Internet Things J **1**(1), 22–32 (2014)

A Network Calculus Based Traceable Performance Analysis Framework of C-RAN for 5G

Muzhou Xiong, Haixin Liu, Deze Zeng, and Lin Gu

1 Introduction

Recent years have witnessed the proliferation of mobile communications market, in both user number and traffic volume. As forecast by Cisco [1], the number of mobile users will increase to 11.6 billion and the mobile traffic volume to 48.3 exabytes per month by 2021. To meet the dramatic traffic volume increase, the next generation mobile networks (5G) are expected to obtain 1000× capacity as compared to current mobile networks. Aiming at the 1000× capacity challenge, much effort has been paid by either improving the density of base stations and antenna (like HetSnet [2], MIMO [3]), or by exploiting more spectrum resources (e.g., millimeter wave [4] as the extra spectrum), due to current mobile networks approaching the Shannon Limit.

The densely deployed base stations with complex construction will definitely lead to high operational expenditure in terms of energy consumption and maintenance. China Mobile estimates that 72% energy is consumed by base station [5] in mobile networks. Towards energy-efficient and low-cost architecture for the next generation mobile network, cloud radio access network (C-RAN) was first proposed by IBM [6] in 2009 and further developed by China Mobile [5], which has been widely considered as a key technique for 5G [7, 8]. Typical C-RAN architecture includes three components, i.e., remote radio head (RRH), base band unit (BBU) pool, and front-haul. RRH, a set of distributed radio units with antenna, realizes the

M. Xiong · H. Liu · D. Zeng (✉)
School of Computer Science, China University of Geosciences, Wuhan, China
e-mail: deze@cug.edu.cn

L. Gu
School of Computer Science and Technology, Huazhong University of Science and Technology, Wuhan, China

B. Ye et al. (eds.), *2nd International Conference on 5G for Ubiquitous Connectivity*, EAI/Springer Innovations in Communication and Computing,
https://doi.org/10.1007/978-3-030-22316-8_13

signal transmission to users. BBU pool, offloads the baseband processing function from base stations to a centralized resource pool, in which dynamic resource allocation and sharing can be implemented aiming at improving energy efficiency and resource utility. As for fronthaul, it connects RRHs and BBU pool, aggregating baseband samples (I/Q data) from RRHs to BBU pool. The benefits of the C-RAN architecture exist in the following aspects. The light-weighted RRHs without baseband processing function can reduce the energy consumption; on the other hand, the energy consumption of BBU pool can be further decreased due to resource sharing and centralized cooling system. In addition, the system maintenance cost also decreases since the baseband processing units are deployed in one site. C-RAN has attracted research interest from both academic [9–11] and industry [12, 13]. For example, Sundaresan et al. [9] propose a flexible and re-configurable front-haul to provide transmission strategies with different mobile traffic patterns. Dahrouj et al. [10] propose a coordinated scheduling, hybrid back-hauling, and multicloud association for C-RAN in heterogeneous environment. Towards energy efficiency, Peng et al. [11] propose a resource assignment and power control to maximize the data transmission volume per Watt for down-link in heterogeneous C-RAN.

Much effort has been devoted to improving C-RAN energy efficiency and resource utilization. Peng et al. [14] propose a closed form expression for ergocic capacity for RRH association by assuming the RRH distribution as two-dimensional Poisson Point Process. Liao et al. [15] formulate a optimization model to minimize the required computing resource with given RRH number and different traffic patterns. On the other hand, we realize that application in mobile network needs different quality of service (QoS) in terms of delay. For example, short text service (SMS) can tolerant a much larger delivery delay than calling service. However, existing work ignores this difference and considers all the requests with the same QoS. By this assumption, more resource will be needed to qualify the unrequired higher QoS. Regarding this, we in this paper propose a performance analysis framework for C-RAN, classifying traffic flows from applications with different priorities in terms of delay tolerance. With the analysis framework, the upper bounds of processing delay can be derived, which can be used to further calculate the required processing capacity in the BBU pool.

Due to the mobility of user device, the mobile traffic offers various patterns in spatio-temporal dimensions. Hence, the performance analysis for C-RAN should be aware of variety of mobile traffic, usually described by stochastic model. For example, in [16] identical independent Piosson distribution is assumed for the traffic from each RRH with the aim of resource allocation optimization in C-RAN. Tang et al. [17] apply a double-layer queueing network to represent each UE's data processing and transmitting behavior for the downlink of C-RAN. However, if the real traffic varies away from the assumed stochastic model, the derived analysis results by those methods deviate from the real value. As alternative deterministic method for classic queuing problem, network calculus [18] has been widely developed as a powerful means for communication system modeling and performance analysis [19, 20]. Aided by network calculus, delay bound can be derived with closed form expressions without relying on any traffic distribution assumption. Regarding this, we in this paper apply network calculus for the performance analysis framework, aiming at obtaining a closed form expression for the delay upper-bound for traffic flows with priority. The framework only needs the mean traffic rate and the maximum burst for the analysis. The mean rate and burst volume of traffic can be easily obtained from historical data trace. With the results from the proposed framework, it can further direct the design and maintenance for C-RAN, determining the required computing capacity in run-time.

To the best of our knowledge, we are the first to propose the performance analysis framework for C-RAN with the application classification in terms of delay tolerance. The contributions of the paper exists in the following two aspects.

- We propose a closed form expression analysis framework for applications in C-RAN with different priorities in terms of delay tolerance.
- Extensive experiments with numerical analysis are conducted, which validates our proposed performance analysis framework.

The rest of the paper is organized as follows. A brief review of the existing work is summaries in Sect. 2. The system models and preliminary of network calculus is given Sect. 3. The performance analysis framework is derived in Sect. 4 and validated in Sect. 5. The paper is concluded in Sect. 6.

2 Related Work

In this section we briefly summarize the state-of-art of C-RAN and its performance analysis.

C-RAN was first proposed by IBM [6] in 2009 and further developed by China Mobile [5]. By decoupling the baseband processing functionality from the conventional base station, the traditional base station is evolved to RRH dedicating in the functionality of signal processing. The baseband processing is then offloaded to a centralized computing pool, i.e., BBU pool. The two main components of C-RAN is connected by front-haul, usually by optical link. Such separation is able to lead advantages including high throughput and energy efficiency. Detailed analysis of C-RAN advantages can be found in the survey [21].

The performance analysis of C-RAN has been conducted in order to optimize either resource allocation or energy consumption with the QoS guarantee for different scenarios. For the aspect of capacity of RRH, a closed form expression for ergocic capacity for RRH association is proposed in [14] for both single and multiple nearest RRH association strategies. In [22], the outage probability and minimal required number of RRHs are derived by assuming the distribution of RRH location as Poisson point process. Yang et al. [23] further analyzes the CRAN performance with uniformly distributed base stations with the considerations of both small-scale Reyleigh fading and large-scale path loss. By using Gauss-Chebyshev integration, the method obtains tightly approximation of outage probability and ergodic rate. Aiming at mitigating the interior interference and improve energy efficiency performances in heterogeneous CRAN, a joint energy-efficient resource assignment and power allocation optimization is proposed in [11] thus to improve the soft fractional frequency reuse. Zhan et al. [24] formulate a formation game to manage interference among RRH, which directs the decision of RHH to serve the mobile devices and gains higher throughput. In order to enhance energy efficiency of C-RAN, an optimization problem is proposed in [11], considering resource assignment and power allocation.

Different from the existing work, this work put the emphasis on building a analysis framework for the C-RAN to obtain delay bound. Aided by network calculus, the proposed framework does not need the knowledge of statistic distributions for mobile traffic and RRH locations. In addition, we consider mobile traffic from different applications with various priorities in terms of delay tolerance, reflecting diverse QoS requirements for mobile applications.

3 System Model and Preliminaries

We consider the C-RAN architecture and the traffic flow queuing model with priorities as illustrated in Fig. 1. An RRH serves a set of mobile devices for signal processing, while the baseband processing for all RRHs is aggregated and offloaded to the centralized BBU pool as in Fig. 1a. RRH and BBU pool are connected by optical fronthaul. We divide applications with three priorities in terms of delay tolerance, i.e., RRH by call service, Internet services, and short message service, with priority from high to low (as shown in Fig. 1b). Higher priority means that the corresponding service requires lower delay. The BBU pool uses the preemptive first-in-first-out policy to schedule the arrived traffic. This means that when the BBU pool receives a traffic request from a higher flow, it interrupts the current computing

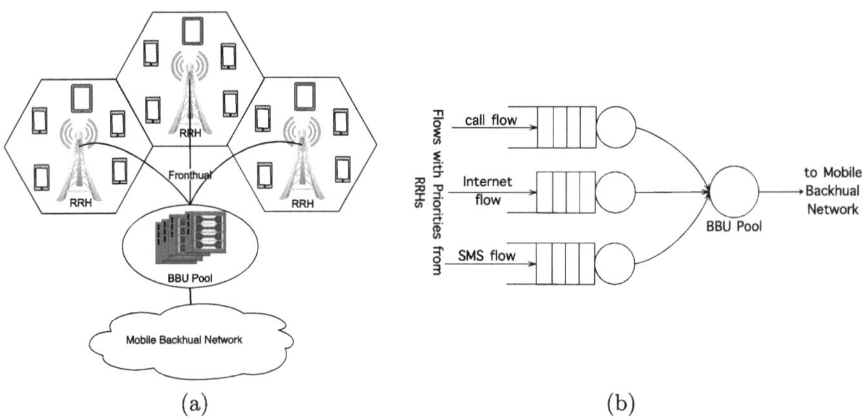

Fig. 1 System model for performance analysis framework of C-RAN. (**a**) Architecture of C-RAN. (**b**) Traffic queueing model with priorities

for the flow with lower priority. Each traffic flow is generated by user device (UE) and aggregated by the associated RRH. We also assume that the BBU pool holds a constant service rate R bit/s.

In the rest of this section, we briefly give several definitions and propositions as the basic of the Network calculus theory. More detailed theorems and proofs can be found in [18, 25].

The network calculus is established based on the min-plus algebra, with the basic operations including min-plus convolution and min-plus deconvolution, defined as follows.

Definition 1 (Min-Plus Convolution) Let f and g be two wide-sense increasing functions. The min-plus convolution of f and g is the function

$$(f \otimes g)(t) = \inf_{0 \leq s \leq t} \{f(t-s) + g(s)\}. \tag{1}$$

Definition 2 (Min-Plus Deconvolution) Let f and g be two wide-sense increasing functions. The min-plus deconvolution of f and g is the function

$$(f \oslash g)(t) = \sup_{u \geq 0} \{f(t+u) - g(u)\}. \tag{2}$$

In order to describe the volume of arrived and processed traffic, we denote $R_i(t)$ and $R_i^*(t)$, two wide-sense increasing functions, used to indicate the input and output traffic volume in the duration $(0, t]$ for different types of flows, where $i \in \{h, m, l\}$ representing call, Internet services, and SMS, respectively. Network calculus uses *Arrival curve* and *service curve* to describe the characteristics of the input and output traffic, respectively.

Definition 3 (Arrival Curve) Given wide-sense increasing function $\alpha_i(t), t \geq 0$, we say that the flow $R_i(t)$ is constrained by α_i if and only if for all $s \leq t$:

$$R_i(t) - R_i(s) \leq \alpha_i(t-s). \tag{3}$$

We say that $R_i(t)$ of flow f_i has $\alpha_i(t)$ as an *arrival curve*.

Definition 4 (Service Curve) Suppose $R_i(t)$ and $R_i^*(t)$ are the wide-sense increasing functions for the input and output traffic volume functions for a flow with arrival curve $\alpha_i(t)$. We say that the BBU pool offers a service curve β_i for flow i if and only if β_i is wide-sense increasing function and $R_i^*(t) \geq R_i(t) \otimes \beta_i(t)$.

Two main conclusions of network calculus are given by the following proposition, i.e., the upper-bound expressions for the delay.

Proposition 1 (Delay Bound) *The delay bound $P(t)$ of a system with arrival curve α and service curve β is expressed as*

$$\begin{aligned} P(t) &\leq h(\alpha, \beta), \\ \text{and} h(\alpha, \beta) &= \sup_{t \geq 0} \{\inf\{d \geq 0 : \alpha(t) \leq \beta(t+d)\}\}. \end{aligned} \tag{4}$$

4 Performance Analysis Framework for C-RAN

Based on Proposition 1, we derive the closed form expressions for the performance analysis framework for C-RAN. The main result of the framework includes the delay upper-bound for applications with priorities in the C-RAN system. We first derive the service curves for the three flows with processing priority from low to high. Based on the derived service curves, we obtain the upper-bounds for the corresponding flows. All the notations used in the following analysis are listed in Table 1. Specially, the subindex for the notations could be h, m, l, representing specified metric for flows with high, medium, and low priority for calling service, Internet service, and SMS, respectively.

4.1 Service Curve for Flows with Different Priorities

We classify mobile applications into three different priorities as call flow, SMS flow, and general flow.Priority is the highest call flow, followed by general traffic, and finally text messages.

In order to use Proposition 1, the arrival and service curves for each of the three flows should be obtained first. We assume the arrival curve for flow i with the following expressions:

$$\alpha_i(t) = r_i(t) + b_i, i \in \{h, m, l\}, \tag{5}$$

where r_i represents the mean rate for flow i, and b_i for the burst traffic volume. The processing capacity of BBU is fixed with the value R. With the assumed arrival curves for flows with different processing priorities, we derive the corresponding service curves as in the following theorems.

Lemma 1 *The service rate of BBU Pool is R and $r_h + r_m + r_l \leq R$. Otherwise, the bounds of the system will approach infinity.*

The lemma can be represented as in Fig. 2. In order to make the analysis practical, we assume the condition always holds.

Table 1 Frequently used notations

f_i	Flows with priority $i, i \in \{h, m, l\}$
$\alpha_i(t)$	Arrival curve of flow with priority i
$\beta_i(t)$	Service curve of flow with priority i
$R_i(t)$	Input and output cumulative function of flow with priority i
$R_i^*(t)$	Cumulative output function for flow with priority i
d_i	Delay bound of flow with priority i

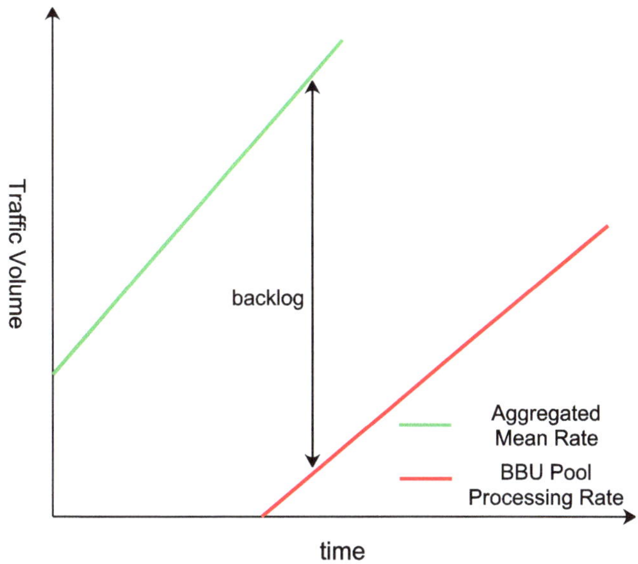

Fig. 2 Rate constraints for C-RAN system with priorities

Theorem 1 (Service Curve of f_l) *The service curve of $f_l(t)$ is*

$$\beta_l(t) = (R - r_h - r_l)(t - \frac{b_h + b_l}{R - r_h - r_m})^+. \tag{6}$$

Proof Suppose at time s, the system begins to backlog. Since the flow f_l is with the lowest priority, the backlog traffic must be from f_l due to the preemptive scheduling policy. The backlogged traffic will be served after traffic from the higher queue is processed. For any time $t, t < s$, the processed traffic volume for flow f_l in the time period $(s, t]$ can be expressed as:

$$\begin{aligned} &R_l^*(t) - R_l^*(s) \\ &= R * (t - s) - (R_h^*(t) - R_h^*(s)) - (R_m^*(t) - R_m^*(s)), \end{aligned} \tag{7}$$

where R is the fixed processing rate of the BBU pool.

Since only the flow f_l is backlogged at time s, and there is no backlog for flows with higher priorities, i.e., f_m and f_h. Hence, we have the following equations for f_m and f_h at time s:

$$\begin{aligned} R_h(s) - R_h^*(s) = 0, \\ R_m(s) - R_m^*(s) = 0. \end{aligned} \tag{8}$$

Hence, we have the following inequality for flow f_h:

$$\begin{aligned} R_h^*(t) - R_h^*(s) =& R_h^*(t) - R_h(s) \\ \leq& R_h(t) - R_h(s) \\ \leq& \alpha_h(t-s). \end{aligned} \tag{9}$$

Similarly, we also have the following inequality for flow f_m:

$$R_l^*(t) - R_l^*(s) \leq \alpha_l(t-s). \tag{10}$$

Substituting (9) and (10) into (7), we can obtain:

$$\begin{aligned} R_l^*(t) - R_l^*(s) \geq& R*(t-s) - \alpha_h(t-s) - \alpha_m(t-s) \\ \geq& (R*(t-s) - \alpha_h(t-s) - \alpha_m(t-s))^+. \end{aligned} \tag{11}$$

Let function $\beta_l(t-s)$ defined as:

$$\beta_l(t-s) = [R*(t-s) - \alpha_h(t-s) - \alpha_m(t-s)]^+, \tag{12}$$

we can rewrite (11) as:

$$R_l^*(t) - R_l^*(s) \geq \beta_l(t-s). \tag{13}$$

With the arbitrary of t, we can further transform the inequality as:

$$\begin{aligned} R_l^*(t) \geq& R_l^*(s) + \beta_l(t-s) \\ \geq& \inf_{0\leq s\leq t}(R_l^*(s) + \beta_l(t-s)) \\ =& (R_l \otimes \beta_l)(t). \end{aligned} \tag{14}$$

Since $R \geq r_m + r_h + r_l$, the function $\beta_l(t)$ is wide-sense increasing. As defined by Definition 4, we can state that the function $\beta_l(t)$ is the service curve of flow f_l. With some transformations, we obtain

$$\begin{aligned} \beta_l(t) =& (R*t - \alpha_h(t) - \alpha_m(t))^+ \\ =& (R*t - (r_h t + b_h) - (r_m t + b_m))^+ \\ =& (R - r_h - r_m)(t - \frac{b_h + b_m}{R - r_h - r_m})^+. \end{aligned} \tag{15}$$

This completes the proof. □

The following two theorems gives the service curves for flows with medium and high processing priorities, respectively. The proof for the two theorems is similar to Theorem 1.

Theorem 2 (Service Curve of Flow f_m) *The service curve of flow f_m is*

$$\beta_m(t) = (R - r_h)(t - \frac{b_h}{R - r_h})^+. \tag{16}$$

Theorem 3 (Service Curve of Flow f_h) *The service curve of $f_h(t)$ is*

$$\beta_h(t) = R * (t - 0)^+ \tag{17}$$

4.2 Delay Upper-Bound Analysis

With the derived service curves, we derive the delay upper-bound for different flow, which indicates the delay value in the worst case.

Theorem 4 (Delay Bound of f_l) *For the lowest priority flow f_l, its delay bound is:*

$$d_l = \frac{b_l + b_m + b_h}{R - r_h - r_m} \tag{18}$$

Proof According to Proposition 1, the delay bound of flow f_l in the worst case is the supremum of the horizontal deviation between arrival curve $\alpha_l(t)$ and service curve $\beta_l(t)$, i.e.,

$$\begin{aligned} d_l(t) &= h(\alpha_l, \beta_l) \\ &= \sup_{t \geq 0}\{\inf\{d \geq 0 : \alpha_l(t) \leq \beta_l(t + d)\}\} \end{aligned} \tag{19}$$

For $\alpha_l(t) \leq \beta_l(t + d)$, using the flow's arrival curve (5) and service curve (6), we have:

$$r_l + b_l \leq (R - l_m - l_h)(t + d - \frac{b_m + b_h}{R - r_m - r_h})^+. \tag{20}$$

With further transformations for (20) as follows, we obtains:

$$\begin{aligned} d &\geq \frac{r_l t + b_l + b_m + b_h}{R - r_m - r_h} - t \\ &\geq \frac{(r_l + r_m + r_h - R)t + b_m + b_h}{R - r_m - r_h}. \end{aligned} \tag{21}$$

With this, (19) can be rewrite as:

$$d_l = \sup_{t\geq 0}\{\inf d \geq 0 : d \geq \frac{(r_l + r_m + r_h - R)t + b_m + b_h}{R - r_m - r_h}\} \tag{22}$$

Since $r_l + r_m + r_h < R$, the function $\frac{(r_l+r_m+r_h-R)t+b_m+b_h}{R-r_m-r_h}$ is strictly decreasing with $t \geq 0$, the maximum value of which is obtained when $t = 0$. This indicates that:

$$\begin{aligned} d_l &= \sup_{t\geq 0}\{\inf\{d \geq 0 : d \geq \frac{(r_l + r_m + r_h - R)t + b_m + b_h}{R - r_m - r_h}\}\} \\ &= \inf\{d \geq 0 : d \geq \frac{b_m + b_h}{R - r_m - r_h}\} \\ &= \frac{b_m + b_h}{R - r_m - r_h}, \end{aligned} \tag{23}$$

which completes the proof. □

The following theorems give the delay bound for the flow with the medium and high processing priorities. The proof is similar to the that of Theorem 4.

Theorem 5 (Delay Bound of f_m) *The delay bound of flow f_m with the medium processing priority is:*

$$d_m(t) = \frac{b_h + b_m}{R - r_h} \tag{24}$$

Theorem 6 (Delay Bound of f_h) *For the flow f_h, its delay bound is:*

$$d_h(t) = \frac{b_h}{R} \tag{25}$$

5 Simulation Results

So far, we have derive the service curve and worst-case upper-bound of delay for C-RAN architecture. In this section, numerical simulations are conducted to validate the proposed performance analysis framework. We fix the delay tolerance for the lowest priority as 5 ms.

For the first numerical simulation, we assume that all flows with the same arrival curve for all the three flows with mean rate as 1500 kb/s and burst traffic volume as 100 kb. Hence, the arrival curve for all the three flows is expresses as

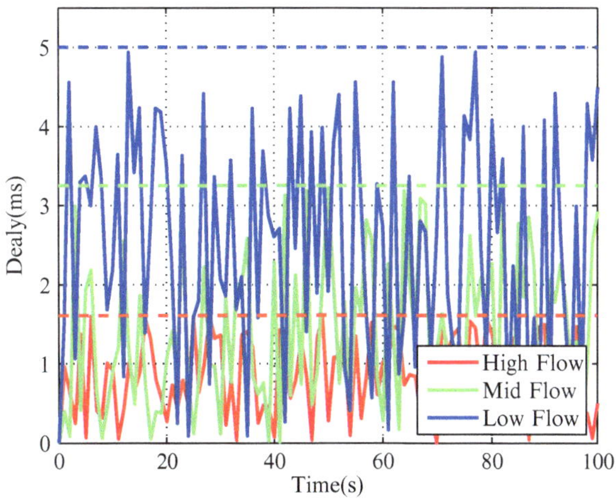

Fig. 3 Simulation results for delay with the same arrival curve

$\alpha_i(t) = 1500t + 100, i \in \{h, m, l\}$. According to theorems given in Sect. 4.2, the delay bounds for the low, medium, and high flows are 5 ms, 3.25 ms, and 1.61 ms, respectively. In this simulation, we randomly generate traffic for each flows strictly constrained by the arrival curves for the flows, and obtains delay for each flow, as illustrated in Fig. 3. The required process rate for the BBU pool can be calculated as 63 Mb/s according to Lemma 1.

We also illustrates the delay for each flow with different priority with dotted lines in these figures. As the simulation results indicate that the obtained delay is less than or equal to the corresponding calculated upper-bounds, since the generated traffic volume is constrained by the arrival curve.

In order to validate the proposed framework in a scenario similar to the real 5G environment, we conduct another set of experiments for flows with different arrival curves. The arrival curves for the flows are defined as in Eq. (26). The units for mean rate and burst tolerance are kb/s and kb, respectively. All the other simulation parameters are the same to the experiments with the same arrival curve in the previous experiment.

According to the proposed analysis framework, we can derive the delay bounds as 1.61 ms, 3.25 ms, and 5 ms with priority from high to low. We illustrate the simulation results as in Fig. 4. The results indicate that delay results are always less than the derived theoretical values (illustrated as the dotted line with the same color as the corresponding flow) for each flow by the framework. This further validate the proposed performance analysis framework for C-RAN.

$$\begin{cases} \alpha_h = 1000t + 100 \\ \alpha_m = 1500t + 150 \\ \alpha_l = 2000t + 200 \end{cases} \tag{26}$$

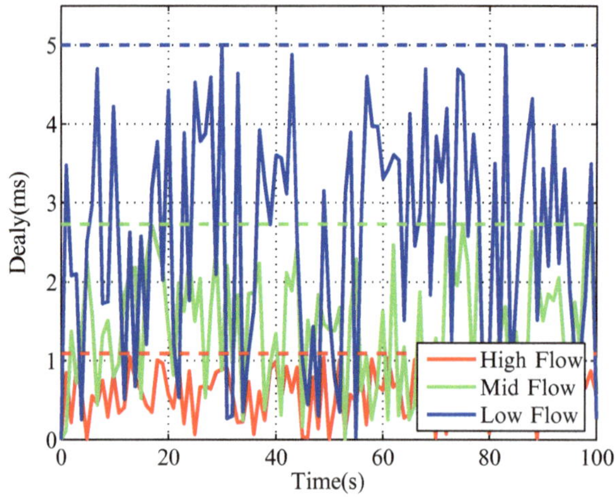

Fig. 4 Simulation results for delay with different arrival curve

6 Conclusions

As C-RAN is widely considered as one of the key enabler for 5G due to its energy efficiency and flexible architecture, it has attracted much research interest from both academic and industry. We in this paper propose a performance analysis framework for C-RAN. Unlike the existing work, the proposed analysis framework applies network calculus and classify applications in terms of different required QoS. The performance analysis process then does not require any stochastic assumptions by applying the theory of network calculus, and can further improve the energy efficiency through distinguishing the required QoS. The experiments results, by numerical simulations and experiments with real data trace, validate the performance analysis framework in different scenarios. The proposed analysis framework can be used to further direct the design and deployment of C-RAN for mobile network operator in 5G.

Acknowledgment This research was supported by the NSF of China (Grant Nos. 61772480, 61602199, 61402425)

References

1. Cisco visual networking index: global mobile data traffic forecast update, 2016–2021, Cisco, Tech. Rep. (2017)
2. Fortes, S., Aguilar-García, A., Barco, R., Barba, F.B., Fernández-Luque, J.A., Fernández-Durán, A.: Management architecture for location-aware self-organizing LTE/LTE-a small cell networks. IEEE Commun. Mag. **53**(1), 294–302 (2015)

3. Choi, L.-U., Murch, R.D.: A transmit preprocessing technique for multiuser mimo systems using a decomposition approach. IEEE Trans. Wirel. Commun. **3**(1), 20–24 (2004)
4. Rappaport, T.S., Sun, S., Mayzus, R., Zhao, H., Azar, Y., Wang, K., Wong, G.N., Schulz, J.K., Samimi, M., Gutierrez Jr., F.: Millimeter wave mobile communications for 5G cellular: it will work! IEEE Access **1**(1), 335–349, 2013
5. C-RAN: the road towards green RAN, China Mobile, Tech. Rep. (2011)
6. Lin, Y., Shao, L., Zhu, Z., Wang, Q., Sabhikhi, R.K.: Wireless network cloud: architecture and system requirements. IBM J. Res. Dev. **54**(1), 4–1 (2010)
7. Li, Y., Jiang, T., Luo, K., Mao, S.: Green heterogeneous cloud radio access networks: potential techniques, performance trade-offs, and challenges. IEEE Commun. Mag. **55**(11), 33–39 (2017)
8. Li, S., Xu, L.D., Zhao, S.: 5G internet of things: a survey. J. Ind. Inf. Integr. **10**, 1–9 (2018). http://www.sciencedirect.com/science/article/pii/S2452414X18300037
9. Sundaresan, K., Arslan, M.Y., Singh, S., Rangarajan, S., Krishnamurthy, S.V.: Fluidnet: a flexible cloud-based radio access network for small cells. IEEE/ACM Trans. Netw. **24**(2), 915–928 (2016)
10. Dahrouj, H., Douik, A., Dhifallah, O., Al-Naffouri, T.Y., Alouini, M.: Resource allocation in heterogeneous cloud radio access networks: advances and challenges. IEEE Wirel. Commun. **22**(3), 66–73 (2015)
11. Peng, M., Zhang, K., Jiang, J., Wang, J., Wang, W.: Energy-efficient resource assignment and power allocation in heterogeneous cloud radio access networks. IEEE Trans. Veh. Technol. **64**(11), 5275–5287 (2015)
12. Huawei: Cloud RAN introduction. In: The 4th CJK International Workshop - Technology Evolution and Spectrum (2011)
13. ZTE green technology innovations white paper, ZET, Tech. Rep. (2011)
14. Peng, M., Yan, S., Poor, H.V.: Ergodic capacity analysis of remote radio head associations in cloud radio access networks. IEEE Wireless Commun. Lett. **3**(4), 365–368 (2014)
15. Liao, Y., Song, L., Li, Y., Zhang, Y.A. How much computing capability is enough to run a cloud radio access network?. IEEE Commun. Lett. **21**(1), 104–107 (2017)
16. Li, J., Peng, M., Cheng, A., Yu, Y., Wang, C.: Resource allocation optimization for delay-sensitive traffic in fronthaul constrained cloud radio access networks. IEEE Syst. J. **11**(4), 2267–2278 (2017)
17. Tang, J., Tay, W.P., Quek, T.Q.S.: Cross-layer resource allocation with elastic service scaling in cloud radio access network. IEEE Trans. Wireless Commun. **14**(9), 5068–5081 (2015)
18. Cruz, R.L.: A calculus for network delay. I. Network elements in isolation. IEEE Trans. Inf. Theory **37**(1), 114–131 (1991)
19. Parekh, A.K., Gallagher, R.G.: A generalized processor sharing approach to flow control in integrated services networks: the multiple node case. IEEE/ACM Trans. Netw. (ToN) **2**(2), 137–150 (1994)
20. Boudec, J.L.: Application of network calculus to guaranteed service networks. IEEE Trans. Inf. Theory **44**(3), 1087–1096 (1998)
21. Checko, A., Christiansen, H.L., Yan, Y., Scolari, L., Kardaras, G., Berger, M.S., Dittmann, L.: Cloud RAN for mobile networksa technology overview. IEEE Commun. Surv. Tutorials **17**(1), 405–426 (2015)
22. Ding, Z., Poor, H.V.: The use of spatially random base stations in cloud radio access networks. IEEE Signal Process. Lett. **20**(11), 1138–1141 (2013)
23. Yang, Z., Ding, Z., Fan, P.: Performance analysis of cloud radio access networks with uniformly distributed base stations. IEEE Trans. Veh. Technol. **65**(1), 472–477 (2016)
24. Zhan, S.C., Niyato, D.: A coalition formation game for remote radio heads cooperation in cloud radio access network. IEEE Trans. Veh. Technol. **PP**(99), 1–1 (2016)
25. Cruz, R.: A calculus for network delay. II. Network analysis. IEEE Trans. Inf. Theory **37**(1), 132–141 (1991)

Image Dehazing Using Degradation Model and Group-Based Sparse Representation

Xin Wang, Xin Zhang, Hangcheng Zhu, Qiong Wang, and Chen Ning

1 Introduction

Images of outdoor scenes captured in haze, fog, mist conditions are easily degraded [1, 2]. The problem of haze removal has thus attracted a large number of researchers. How to remove fog/haze from an image has been extensively studied, and decades of research on this topic have produced a diverse set of approaches [3–14].

From different views, these algorithms have two types. One is thought as image enhancement from the view of image processing [3–9]. Although various methods via image enhancement theory have been proposed to dehaze images, they hardly obtain high-quality results, for haze degradation actually has a close relationship with the physical characteristics of haze, which the above mentioned methods do not take into account. The restoration algorithm based on physical models is another class of methods, where the physical characteristics of haze are modeled [10–14]. Haze-free images can be restored accurately using these models. For image

X. Wang (✉)
College of Computer and Information, Hohai University, Nanjing, China

Jiangsu Key Laboratory of Image and Video Understanding for Social Safety, Nanjing University of Science and Technology, Nanjing, China
e-mail: wang_xin@hhu.edu.cn

X. Zhang · H. Zhu
College of Computer and Information, Hohai University, Nanjing, China

Q. Wang
Jiangsu Key Laboratory of Image and Video Understanding for Social Safety, Nanjing University of Science and Technology, Nanjing, China
e-mail: wang_xin@hhu.edu.cn

C. Ning
School of Physics and Technology, Nanjing Normal University, Nanjing, China

B. Ye et al. (eds.), *2nd International Conference on 5G for Ubiquitous Connectivity*, EAI/Springer Innovations in Communication and Computing,
https://doi.org/10.1007/978-3-030-22316-8_14

dehazing algorithms based on physical models, since image prior knowledge such as depth information or atmospheric conditions plays a critical role in the models, it makes such methods quite inconvenient and easy to fail in the cases where the estimation of related information is imprecise.

Recently, sparse representation (SR) has come out as a strong tool for digital image denoising [15], deblurring [16], and super-resolution [17]. For image restoration, group-based sparse representation, i.e., GSR, put forward by Zhang *et al.* in [18] have become very popular recently. Inspired by the success application of GSR to image restoration, this paper extends it to image dehazing problem. However, so as to effectively utilize the framework of GSR to process the ill-posed nature of dehazing, it is necessary to construct a suitable degradation model at first.

Therefore, this study proposed a method based on the classical dichromatic atmospheric scattering model to develop a novel degradation model in the first place. Then, an integration of the degradation model and GSR method is utilized to cope with the image dehazing problem. Ultimately, the proposed image dehzing method exhibits two advantages. First, degradation model constructed via the dichromatic atmospheric scattering model effectively explores the physical characteristics of hazy images, which makes the proposed method physically sound. Second, a novel framework for image dehazing is established based on group-based sparse representation. As a result, the framework may produce impressive restoration results.

2 Preliminaries

The objective of restoration is to recover the high quality image x from its observed degraded image y. This ill-posed inverse problem can be modeled by [19]:

$$y = Hx + \eta \tag{1}$$

where H denotes the degradation operator. η represents an additive noise, which is usually considered as additive Gaussian white noise. The purpose is to obtain an estimate of the original image. Obviously, the more we know about H, the closer the estimate will be to the original image.

In view of the importance of H in image restoration, we propose to generate the degradation operator H inspired by a classical physical model, that is, the dichromatic atmospheric scattering model in our proposed algorithm.

3 Presented Approach

The presented algorithm consists of two main parts: degradation model construction and image dehazing via degradation model and group-based sparse representation.

3.1 Degradation Model Construction

In a color hazy image, the scene point **F**'s color can be described as a vector combination of the color of airlight ($m\hat{\mathbf{A}}$) and the color of clear day ($n\hat{\mathbf{C}}$) [20]:

$$\mathbf{F} = m\hat{\mathbf{A}} + n\hat{\mathbf{C}} \tag{2}$$

$$m = F_{\infty}\left(1 - e^{-\beta d}\right) \tag{3}$$

$$n = \mathrm{Re}^{-\beta d} \tag{4}$$

where $\hat{\mathbf{A}}$ and m denote the direction and amplitude of airlight color. $\hat{\mathbf{C}}$ and n is the direction and amplitude of clear day color. F_{∞} denotes the brightness of sky. R represents the point radiance of clear day. β denotes the atmosphere scattering coefficient. d represents the point depth, which is always estimated as [12]:

$$d = \delta\left(d_{\max} - d_{\min}\right) + d_{\min} \tag{5}$$

where $d_{\max}$ is scene point's maximal depth, and $d_{\min}$ is scene point's minimal depth. $\delta \in [0, 1]$ is a factor for adjustment.

According to the dichromatic atmospheric scattering model as well as the above heuristic depths, we discover the law that the image degradation degree is related to scene point depth. Inspired by this, we propose a scheme to design the degradation operator for hazy images, and based on such operator, the degradation model can be constructed.

The scene point's degradation degree is inversely related to its distance to vanishing point in the image. Based on this, we first set the approximate location of the vanishing point in the image. For simplicity, we assume that the depth of the central position of the image is the largest and regard the central pixel as the vanishing point.

Then, we make the central position as coordinate origin. So the depth of each scene point whose coordinate position is (r_1, r_2) is defined as:

$$d = \frac{1}{r_1^2 + r_2^2 + 1} \tag{6}$$

Based on Eq. 6, we can obtain the depth of the central pixel, that is, $d = 1$.

Third, inspired by the dichromatic atmospheric scattering model, we design the degradation operator for hazy images as follows:

$$H = 1 - e^{-\beta d} = 1 - e^{-\frac{\beta}{r_1^2 + r_2^2 + 1}} \tag{7}$$

where β in [0, 1] is a parameter.

Finally, as we know, degradation operator can be also referred to as the point spread function (PSF). Based on our designed PSF for hazy images, the degradation model can be finally constructed as follows:

$$y = \left(1 - e^{-\beta d}\right) x + \eta \tag{8}$$

By integrating this model into the group-based sparse representation framework, we can estimate x from y.

3.2 *Image Dehazing via Degradation Model and Group-Based Sparse Representation*

Traditional sparse representation uses patch as basic unit [21–23]. Each image patch is supposed to be independent [24]. It neglects self-similarity. As indicated in [18], both self-similarity and sparsity are the important characteristics of natural images, and combining them together can achieve better performance. Based on this idea, group-based sparse representation was proposed by using group rather than patch [18].

In view of the advantages of GSR, in this section, we apply it to solving the single image dehazing problem. The degradation model designed in the above section will be integrated into the GSR framework. The detailed process is described as follows.

First, given an input degraded image y, construct the group for it. Specifically, divide y into n overlapped patches, and each patch is represented by a vector $y_k \in \mathrm{R}^B$ $(k = 1, 2, \ldots, n)$. Find c patches that are best matched for a certain patch in its neighbor area by using Euclidean distance, and then all the similar patches are stacked as $y_{G_k} \in \mathrm{R}^{B \times c}$. This is just called a group.

Second, by integrating the designed degradation model into the group-based sparse representation framework, the image dehazing problem can be written by:

$$\begin{aligned} \hat{\alpha}_G &= \arg\min_{\alpha_G} \tfrac{1}{2} \| H D_G \circ \alpha_G - y \|_2^2 + \lambda \|\alpha_G\|_0 \\ &= \arg\min_{\alpha_G} \tfrac{1}{2} \left\| \left(1 - e^{-\beta d}\right) D_G \circ \alpha_G - y \right\|_2^2 + \lambda \|\alpha_G\|_0 \end{aligned} \tag{9}$$

where $\frac{1}{2} \| H D_G \circ \alpha_G - y \|_2^2$ denotes ℓ_2 term. $\lambda \|\alpha_G\|_0$ represents regularization term and λ is regularization parameter. D_G is a concatenation of $D_{G_k} (k = 1, 2, \ldots, n)$, and D_{G_k} is dictionary self-adaptively learned for group y_{G_k}. α_G is a concatenation of $\alpha_{G_k}(k = 1, 2, \ldots, n)$, and α_{G_k} is sparse coefficient for group y_{G_k}.

Third, adaptive dictionary D_{G_k} for y_{G_k} is learned from the estimate r_{G_k}. We use singular value decomposition to get dictionary atom $d_{G_k \otimes i} \in \mathrm{R}^{B \times c}(i = 1, 2, \ldots, m)$. Ultimately, the adaptively learned dictionary for y_{G_k} can be obtained by:

$$D_{G_k} = \left[d_{G_k \otimes 1}, d_{G_k \otimes 2}, \cdots, d_{G_k \otimes m}\right] \tag{10}$$

Fourth, given D_{G_k}, the sparse coding problem of y_{G_k} over D_{G_k} is to calculate α_{G_k} to make $y_{G_k} \approx D_{G_k}\alpha_{G_k}$. In Eq. 9, image restoration is formulated as the GSR-driven ℓ_0 minimization problem.

Finally, after obtaining the adaptive dictionary D_G and sparse coefficient α_G, the dehazed image is computed as:

$$\hat{x} = D_G \circ \alpha_G \tag{11}$$

4 Experimental Results

4.1 Evaluation Criteria

Here, we conduct a series of experiments on a set of hazy images. Besides qualitative evaluation, in our experiment, we also make the quantitatively evaluation by using three criteria: contrast (C) [25], average gradient (AG) [26, 27], and percentage of number of saturated pixels (PS) [28].

4.2 Qualitative Evaluation

First, we compare the proposed algorithm with four existing algorithms, i.e., Tarel's approach [8], Wang's scheme [12], He's algorithm [13], and Zhang's technique [18]. Some qualitative comparison results are given in Figs. 1, 2, and 3, where the input hazy images are shown in (a), and the dehazed results of our, Tarel's, Wang's, He's, and Zhang's methods are demonstrated in (b)–(f), respectively.

From these figures we can see that, compare with the existing methods, our algorithm achieves the best performance, for it improves the visibility apparently, and the details are well unveiled. Besides, the restored results of our method also have a natural color. On the contrary, the other four algorithms get inferior results. For instance, there is a color distortion in the restored results of Tarel's or He's methods. They make the restored images too dark and at the same time, over-saturated. Even worse, some details are lost in the distance and halo effects are apparent near the depth discontinuities. In Wang's results, neither visibility nor

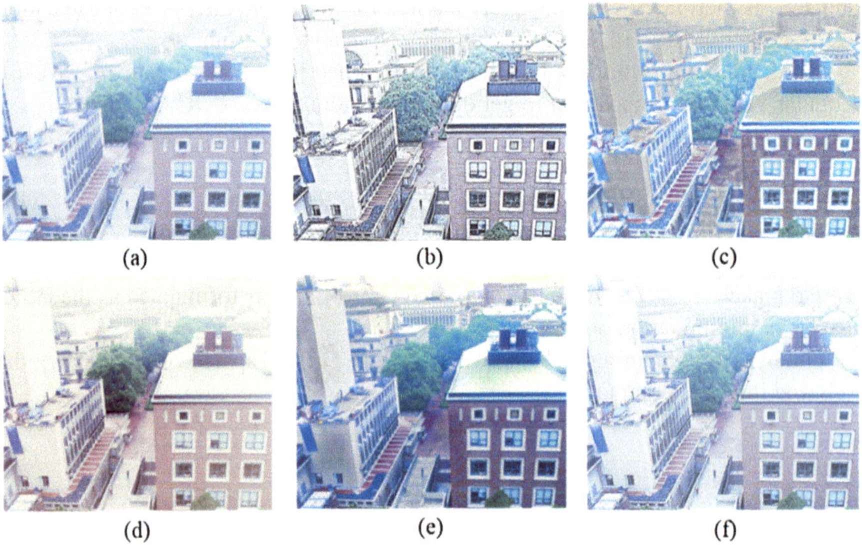

Fig. 1 Example 1: dehazed results with various approaches. (**a**) Original hazy image. (**b**) Our result. (**c**) Tarel's result. (**d**) Wang's result. (**e**) He's result. (**f**) Zhang's result

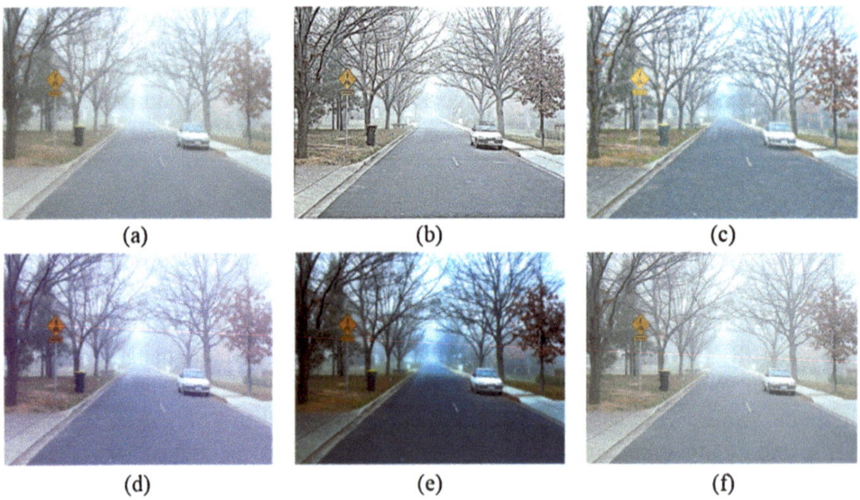

Fig. 2 Example 2: dehazed results with various approaches. (**a**) Original hazy image. (**b**) Our result. (**c**) Tarel's result. (**d**) Wang's result. (**e**) He's result. (**f**) Zhang's result

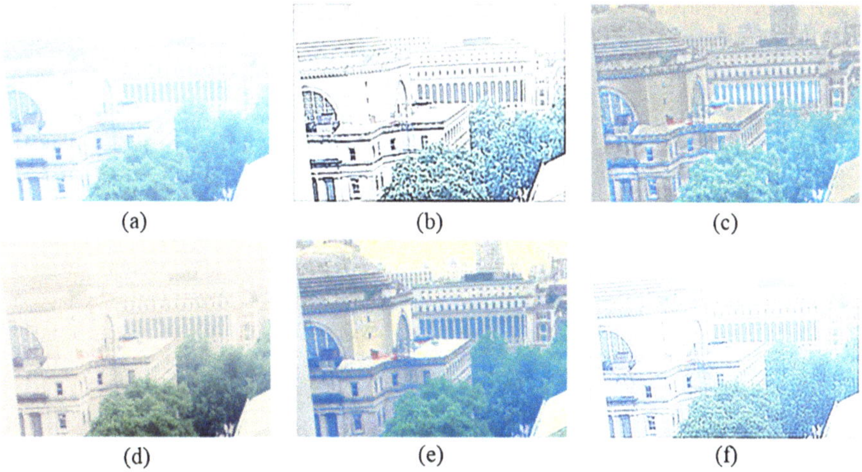

Fig. 3 Example 3: dehazed results with various approaches. (a) Original hazy image. (b) Our result. (c) Tarel's result. (d) Wang's result. (e) He's result. (f) Zhang's result

clarity increases obviously. Thus, its results are unsatisfactory. For Zhang's method, since it only uses the group-based sparse representation strategy for dehazing without suitable degradation model, it also cannot obtain very clear results.

4.3 Quantitative Evaluation

Quantitative results are reported in Table 1. For Figs. 1 and 3, contrast (C) values of our method are highest, while for Fig. 2, the C values of our method are a littler lower than those of He's. It verifies that the dehazed images with the addressed acheme have better visibility. Also, Table 1 gives the average gradient (AG) values for various figures. Note that, compared with other four algorithms, our method achieves highest AG values, which suggests that our results do have greatest clarity. Based on the above, we can draw the conclusion that in the aspect of visibility improvement, our results are comparable to He's. However, our results are much better in regard to clarity. Furthermore, PS descriptor is also computed. Comparison results are given in the last column in Table 1. As expect, our method gets the lowest values of PS. This evidence further proves that the proposed method outperforms other existing algorithms.

Table 1 Quantitative results with various approaches

Indictor	Method	C	AG	PS
Figure 1	Input (a)	0.1144	5.6825	/
	Ours (b)	**0.2275**	**26.2613**	**0.0029**
	Tarel's (c)	0.1403	11.1942	0.0034
	Wang's (d)	0.1369	6.5772	0.0041
	He's (e)	0.2192	10.1583	0.0037
	Zhang's (f)	0.1309	15.1615	0.0031
Figure 2	Input (a)	0.1420	5.7717	/
	Ours (b)	0.2705	**30.4951**	**0.0033**
	Tarel's (c)	0.1686	11.0182	0.0035
	Wang's (d)	0.1707	6.3341	0.0046
	He's (e)	**0.3638**	9.7117	0.0041
	Zhang's (f)	0.1781	15.7396	0.0037
Figure 3	Input (a)	0.0849	4.0436	/
	Ours (b)	**0.1702**	**23.0684**	**0.0072**
	Tarel's (c)	0.1176	10.6665	0.0046
	Wang's (d)	0.1321	5.3178	0.0079
	He's (e)	0.1659	10.2986	0.0052
	Zhang's (f)	0.1110	11.6795	0.0073

Bold indicates the best performance among the comparing methods for each figure

5 Conclusion

An image haze removal algorithm based on degradation model and group-based sparse representation is proposed. We construct the degradation model using the classical dichromatic atmospheric scattering model, and recover the image via GSR. Ultimately, the hybrid degradation model and GSR developed in the work effectively solve hazy images. Experimental results demonstrate that our algorithm outperforms some other methods.

Acknowledgement This work was supported by the Fundamental Research Funds for the Central Universities (2019B15314, 30918014107), National Natural Science Foundation of China (61603124), Six Talents Peak Project of Jiangsu Province (XYDXX-007), 333 High-Level Talent Training Program of Jiangsu Province, and Jiangsu Province Government Scholarship for Studying Abroad.

References

1. Xie, C.H., Qiao, W.W., Liu, Z.: Single image dehazing using kernel regression model and dark channel prior. Signal Image Video Process. **11**, 1–8 (2016)
2. Tripathi, A.K., Mukhopadhyay, S.: Efficient fog removal from video. Signal Image Video Process. **8**(8), 1431–1439 (2014)

3. Kim, T.K., Paik, J.K., Kang, B.S.: Contrast enhancement system using spatially adaptive histogram equalization with temporal filtering. IEEE Trans Consum Electron. **44**(1), 82–87 (1998)
4. Al-Sammaraie, M.F.: Contrast enhancement of roads images with foggy scenes based on histogram equalization. In: Proceedings of the 10th International Conference on Computer Science & Education, pp. 95–101 (2015)
5. Zhou, J., Zhou, F.: Single image dehazing motivated by retinex theory. In: Proceedings of the 2nd International Symposium Instrument Measurement Sensors Net Automation, pp. 243–247 (2013)
6. Mei, X., Yang, J., Zhang, Y., Li, W., Zhang, J.: Video image dehazing algorithm based on multi-scale retinex with color restoration. In: Proceedings of the International Conference Smart Grid Electrical Automation, pp. 195–200 (2016)
7. Grewe, L.L., R, R.: Brooks. Atmospheric attenuation reduction through multisensor fusion. Proc. SPIE. **3376**, 102–109 (1998)
8. Tarel, J.P., Hautière, N.: Fast visibility restoration from a single color or gray level image. In: Proceedings of IEEE International Conference on Computer Vision, pp. 2201–2208 (2009)
9. Tarel, J.P., Hautière, N., Cord, A., Gruyer, D.: Improved visibility of road scene images under heterogeneous fog. Proc. IEEE Intell. Veh. Symp. **23**(3), 478–485 (2010)
10. Oakley, J.P., Satherley, B.L.: Improving image quality in poor visibility conditions using a physical model for contrast degradation. IEEE Trans. Image Process. **7**(2), 167–179 (1998)
11. Narasimhan, S.G., Nayar, S.K.: Interactive (de)weathering of an image using physical models. In: Proceedings ICCV Workshop Color Photometric Methods in Computer Vision, pp. 1–8 (2003)
12. Wang, X., Tang, Z.: Automatic image de-weathering using physical model and maximum entropy. In: Proceedings of the IEEE Conference Cybernetics Intelligent Systems, pp. 996–1001 (2008)
13. He, K., Sun, J., Tang, X.: Single image haze removal using dark channel prior. IEEE Trans. Pattern Anal. Mach. Intell. **33**(12), 2341–2353 (2011)
14. Lai, Y., Chen, Y., Chiou, C., Hsu, C.: Single-image dehazing via optimal transmission map under scene priors. IEEE Trans. Circ. Syst. Video. Technol. **25**(1), 1–14 (2015)
15. Elad, M., Aharon, M.: Image denosing via sparse and redundant representations over learned dictionaries. IEEE Trans. Image Process. **15**(12), 3736–3745 (2006)
16. Dong, W., Zhang, L., Shi, G., Wu, X.: Image deblurring and superresolution by adaptive sparse domain selection and adaptive regularization. IEEE Trans. Image Process. **20**(7), 1838–1857 (2011)
17. Yang, J., Wright, J., Huang, T., Ma, Y.: Image super-resolution via sparse representation. IEEE Trans. Image Process. **19**(11), 2861–2873 (2010)
18. Zhang, J., Zhao, D., Gao, W.: Group-based sparse representation for image restoration. IEEE Trans. Image Process. **23**(8), 3336–3351 (2014)
19. Banham, M.R., Katsaggelos, A.K.: Digital image restoration. IEEE Signal Process. Mag. **14**, 24–41 (1997)
20. Nayar, S.K., Narasimhan, S.G.: Vision in bad weather. In: Proceedings of the IEEE International Conference on Computer Vision, pp. 820–827 (1999)
21. Li, X.: Image recovery from hybrid sparse representation: A deterministic annealing approach. IEEE J. Sel. Top. Signal Process. **5**(5), 953–962 (2011)
22. Mairal, J., Bach, F., Ponce, J., Sapiro, G., Zisserman, A.: Non-localsparse models for image restoration. In: Proceedings of the 12th International Conference on Computer Vision, pp. 2272–2279 (2009)
23. Dong, W., Zhang, L., Shi, G., Li, X.: Nonlocally centralized sparse representation for image restoration. IEEE Trans. Image Process. **22**(4), 1620–1630 (2013)
24. Aharon, M., Elad, M., Bruckstein, A.: K-SVD: An algorithm fordesigning overcomplete dictionaries for sparse representation. IEEE Trans. Signal Process. **54**(11), 4311–4322 (2006)
25. Schechner, Y.Y., Averbuch, Y.: Regularized image recovery in scattering media. IEEE Trans. Pattern Anal. Mach. Intell. **29**(9), 1655–1660 (2007)

26. Zhang, L., Zhang, L., Mou, X., Zhang, D.: FSIM: A feature similarity index for image quality assessment. IEEE Trans. Image Process. **20**(8), 2378–2386 (2011)
27. Wang, J., He, N., Zhang, L., Lu, K.: Single image dehazing with a physical model and dark channel prior. Neurocomputing. **149**, 718–728 (2015)
28. Tripathi, A.K., Mukhopadhyay, S.: Single image fog removal using anisotropic diffusion. IET Image Process. **6**(7), 966–975 (2012)

Delay Analysis for URLLC in 5G Based on Stochastic Network Calculus

Shengcheng Ma, Xin Chen, Zhuo Li, and Ying Chen

1 Introduction

The 5G era is getting closer to us. 5G communication technology appeared for the first time with the 2018 Pyeongchang Winter Olympics in South Korea. It helps audiences watch the live broadcast continuously and smoothly. According to the 5G standard schedule announced by the International Telecommunication Union (ITU), 5G will begin commercialization in 2020 [1]. 5G wireless networks are designed to support diverse and complicated scenarios. The third generation partnership project (3GPP) classify these different scenarios into three big categories: enhanced mobile broadband (eMBB), massive machine type communications (mMTC), and ultra-reliable low-latency communications (URLLC) [2].

URLLC is widely used in self-driving, mission-critical application and some delay sensitive systems. It has stringent requirements in terms of delay and reliability in the 5G New Radio (NR) systems. The key requirements of URLLC as claimed by the 3GPP are to ensure the latency of user plane data less than 1 ms for downlink and uplink, meanwhile to keep very high packet reception reliability about 99.999% [3]. The stringent delay requirement needs new 5G NR technology to bridge the gap. Although existing LTE networks can achieve reliability goals, the cost is some dozens of milliseconds of time delay. That is far away from the criteria of URLLC. So the delay becomes the bottleneck and it needs to be solved. Many academies and companies have proposed some engineering solutions to minimize the delay. Such as the HARQ retransmission or grant-free technology. However, how to analyze the generation of delay from a theoretical perspective and propose a strategy to reduce the delay effectively is an important research subject.

S. Ma · X. Chen · Z. Li (✉) · Y. Chen
Beijing Information Science and Technology University, Beijing, China
e-mail: lizhuo@bistu.edu.cn

B. Ye et al. (eds.), *2nd International Conference on 5G for Ubiquitous Connectivity*, EAI/Springer Innovations in Communication and Computing,
https://doi.org/10.1007/978-3-030-22316-8_15

Stochastic network calculus (SNC) theory is an effective tool to analyze the delay performance. The SNC is a continuous development method to analyze network traffic characteristic and evaluate performance [4]. Different from queuing theory, the SNC permits some packets to violate the desired performance. This feature can better take advantage of statistical multiplexing gains [5]. To deal with random service and statistical guarantee, the SNC theory comes into being with a large number of stochastic processes and network traffic models. Under a suitable traffic model and a chosen server model, the SNC theory can process service guarantee analysis of communication network, such as delay and backlog. So we capitalize on the SNC method to analyze the delay of the 5G URLLC transmission in this paper.

We use stochastic arrival curve to describe the process that user equipment (UE) data sends to gNodeB (gNB) side. According to the 5G network topology architecture, we can deduce the rest stages of data transmission from gNB to cloud server. Every stage of stochastic arrival curve characterizes the delay property, therefore the whole delay of URLLC system is comprised by delays which generated from UE to cloud server.

Our main contributions of this paper can be summarized as follows:

1. We build a tandem model to simulate 5G network architecture. In this model, we can analog the data transmission from UE to cloud server. We use stochastic service process and concatenation property to analysis the latency.
2. Our analysis results represent which parameters are the key factors affecting the delay. By adjusting the key factors, we give some strategies which can reduce the delay effectively.
3. Delay analysis and strategy for reducing latency have valuable theoretical guidance for the design of URLLC deployment. In order to meet stringent delay requirements, it provides guidelines for how to allocate resources.

The rest of this paper is organized as follows. Section 2 summarizes related work of URLLC technology and stochastic network calculus. We present a tandem network model to describe URLLC in Sect. 3. In particular, we illustrate the architecture of this system and analyze the causes of the delay in this section. In Sect. 4, we introduce the experimental environment and analyze the relationship between latency and main factors. We conclude this paper in Sect. 5. Some theoretical proofs are given in Appendix.

2 Related Work

Because the standard of URLLC has not been worked out, many researchers have put forward different solutions for the design of URLLC.

Dozens of researches are focus on how to design and implement URLLC to meet the performance requirements. A design without intervention in the baseband/PHY layer for URLLC is to use interface diversity and integrate multiple communication interfaces. Jimmy and his colleagues propose an analytical framework that combines traditional reliability models with technology-specific latency probability distributions [6]. In this way, they can estimate the performance in terms of latency and reliability in such an integrated communication system. To guarantee a low end-to-end delay with low jitter over combined internet and wireless interfaces, the article [7] presents a new round trip time (RTT) skew control controller with multiple-input multiple-output(MIMO). This controller's advantage is that it solves the data flow split problem at the controlling node. Jaya Rao and Sophie Vrzic have proposed an approach to adopt packet duplication (PD) method to satisfy the latency and reliability requirements [8]. PD technology generates multiple instances and sends them simultaneously in multiple unrelated channels. The receiver selects the best packet according to the channel condition in order to achieve better transmission reliability. This PD technique can provide a cost-effective solution without increasing the complexity in the radio access network (RAN).

In terms of resource allocation and energy efficiency, there are also some researches on URLLC. How the frequency resource will be allocated to the user to send data in URLLC scenario. That is an interesting study which is plunged by Anand and De Veciana [9]. Based on the 5G standard technology Orthogonal Frequency Division Multiple Access (OFDMA), they build a One Shot Transmission model. Adopting queuing theory analysis, they find out a result that a small bandwidth over a longer duration is better than a large swath of bandwidth for short duration in One Shot Transmission system. Green energy saving is getting more and more attention. The article [10] provides a coordinated on-off switching scheme across a set of adjacent gNBs. The gNBs share a sleep schedule among themselves. If the gNBs have lower traffic and fewer connected UEs, they will be set to the OFF mode. This On-Off mode is more energy-efficient than traditional mode on the premise of guaranteeing the time delay.

Because URLLC has strict requirements for delay and reliability, it is very meaningful to evaluate the performance of URLLC. Joachim et al. provide an achievable latency bound evaluation in their article [11]. They compare the latencies for different configurations in 5G RAN transmission. The configuration contains FDD, TDD, frequency numerologies and usage of slots. According to the analysis, a frequency with higher numerology can be used to reduce the latency. An article derived from HUAWEI company proposes a grant-free mode uplink transmission mechanism [12]. Grant-free transmission grant dynamically without scheduling request. This mode is poised to meet the reliability requirement of URLLC in uplink transmission. By simulating random arriving from different numbers of active UEs, the reliability can be improved after adopting the grant-free mode with increasing retransmission.

In order to satisfy the key requirements including latency and reliability, some state-of-the-art solutions have been discussed in [13–16]. These technologies contain fast HARQ retransmission, MIMO, beam forming, diversity interfaces, D2D communication, Ultra Density Network and so on. Some of these technologies can be employed alone to promote the performance, and others need to be combined together to achieve better results. They all mentioned the design of frame structures.

That because low latency and high reliability are contradictory. This requires more flexible frequency and time division.

Stochastic network calculus is a very practical tool, and it has a good practical effect in performance analysis and theoretical boundary calculation. Li and Jiang [17] analyzed the throughput performance on the wireless-powered communication system. They considered the delay as a constrained condition. The stochastic traffic arrival was adopted to derive the cumulative data transmission capacity. Fidler and his partner [18] used stochastic service process to analyze the end-to-end delay performance of TCP. The estimation method of closed-loop flow was implemented by random service process in this paper. They considered both backlog and delay by using stochastic network calculus. In article [19], the performance of wireless finite-state Markov channel was analyzed. The delay boundary was derived based on the moment generating functions (MGF). Xin Chen's team was focused on LTE network and researched the resource allocation to guarantee the delay performance [20]. The delay was constrained by the difference value between stochastic arrival curve and stochastic service curve. The article [21] is an article of cooperation between Fidler M and Jiang. It mainly applies SNC theory to analyze the delay boundary of multi-server systems.

3 System Model

3.1 URLLC Network Architecture

Generally speaking, 5G network includes 5G Standalone networking (SA) and Non-standalone networking (NSA). The SA mode is the establishment of whole new 5G network, while the NSA mode is the combination of 5G and 4G LTE. According to the implementation technology of core network, the NSA mode can continue to be divided. In order to simplify the system model, we only discuss the case of 5G standalone network.

We consider URLLC network as a concatenate system from UE to Cloud. The 4G LTE network is composed by UE, RRU (Radio Remote Unit), BBU (Building Baseband Unit), the EPC (Evolved Packet Core) which is the LTE's core network, and end by cloud servers. Different from 4G LTE, 5G networks are composed by UE, gNB, NGC and Cloud. The gNB contains three parts that are AAU (Active Antenna Unit), DU (Distributed Unit) and CU (Centralized Unit). AAU takes the place of the original RRU and combines some physical layer processing functions of BBU. The BBU function of 4G will be rebuilt into DU and CU in 5G. CU provides the service convergence function in the access side. It focuses on the low real-time capabilities of the protocol stack and adopts a centralized deployment. DU mainly provides data access function to the terminal, including radio frequency and partial signal processing. DU concentrate on the high real-time capabilities of the transport requirements and suit for a distributed deployment method. The NGC (Next Generation Core Network) as Core Network in 5G replace the EPC. 5G NGC

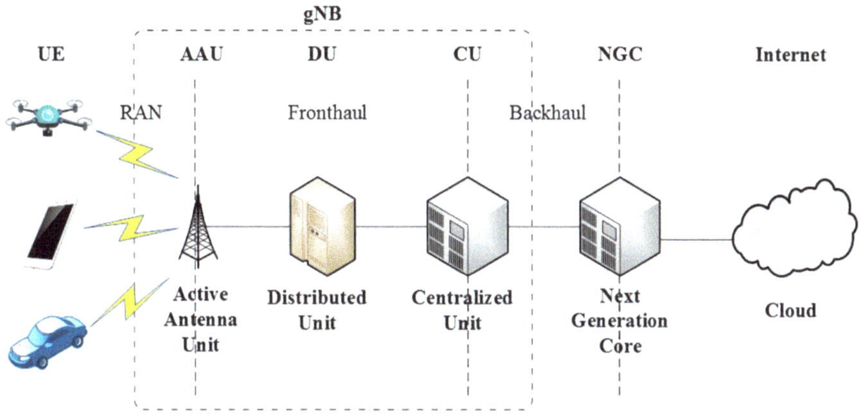

Fig. 1 5G network architecture

is based on SDN/NFV technology and designed to better fit the cloud platform. The architecture is depicted as Fig. 1.

3.2 *Delay of URLLC Network*

UE devices firstly access to AAU. The AAU is actually a part of the base station. This part of the communication belongs to RAN. UE's data will be accepted by AAU, and AAU put forward the data to DU. There are two situations when data arrive at DU. If CU and DU are deployed together, the data can arrive at CU immediately. Otherwise, the data will be sent to CU from DU. The communication from AAU to CU belongs to fronthaul. The data leave CU and continue upward to the NGC. This part of the communication is called backhaul. NGC will process the data and it will take some time. Finally, NGC sends the data to Cloud servers. The unidirectional transmission is finished.

So the whole delay or latency in 5G system is contributed by the time processing of RAN, fronthual, backhaul, Core Network, and Cloud server. It can be expressed as (1).

$$T_{Total} = T_{RAN} + T_{Fronthaul} + T_{Backhaul} + T_{NGC} + T_{Cloud} \tag{1}$$

where

- T_{RAN} is the time cost by physical layer transmission between UEs and AAU.
- $T_{Fronthaul}$ is the delay between AAU to CU. It is the time taken in gNB.
- $T_{Backhaul}$ is the time taken to communication between gNB to NGC.
- T_{NGC} is the delay taken place in NGC.
- T_{Cloud} is the latency which data transmission between NGC and Cloud server.

To meet the URLLC key requirement, we should do a good job of studying T_{Total}. User Plane (UP) latency is the communication time between UE and network nodes when transmission and reception of the data at the corresponding IP layer. Control Plane (C-Plane) latency is the time spend on radio resource allocating and state switching from idle to active. According to the 3GPP acclaims, the C-Plane delay should be less than 10 ms. On the contrary, the User Plane delay should be less than 1 ms. In this part, we discuss the latency mainly on the UP rather than the C-Plane. It because the delay requirement of user plane is higher than that of the control plane.

3.3 Problem Description

The data is transferred from UE, through each node, and finally to the cloud. According to the requirement of URLLC, the reliability and random latency [22] can be described as (2):

$$P\{delay > d\} < \epsilon \tag{2}$$

The delay should be within 1 ms, so the $d = 1$, the unit is millisecond (ms). Where ϵ is defined as a very small probability. The Formula (2) represents the 5G URLLC network successfully transport data and satisfy the delay requirements.

3.4 Stochastic Network Calculus

In SNC theory, the min-plus algebra is applied to analyze queuing system. Let $\mathcal{F}$ denotes the set of non-negative non-decreasing functions and $\bar{\mathcal{F}}$ denotes the set of non-negative non-increasing function. We employ the cumulative process to represent amount of traffic flow. Arrival process, departure process and service process are denoted as $A(t)$, $D(t)$, and $S(t)$ respectively. For any $0 \leq s \leq t$, $A(0) = 0$, $A(s,t) = A(t) - A(s)$, and practical significance of $A(t)$ is the cumulative arrival data at time t. It is the same for $D(t)$ and $S(t)$. Some fundamental definitions of curve are well described in the literature [4]. We utilize and expand the following in this paper.

Definition 1 (Stochastic Arrival Curve) A flow is said to have a stochastic arrival curve $\alpha \in \mathcal{F}$ with bounding function $f \in \bar{\mathcal{F}}$, denoted by $A(t) \sim< f, \alpha >$, if for all $t \geq 0$ and all $x \geq 0$ there holds

$$P\left\{\sup_{0 \leq s \leq t} \{A(s,t) - \alpha(t-s)\} > x\right\} \leq f(x). \tag{3}$$

Where $\alpha(\tau)$ is the stochastic arrival curve, and it denotes the maximum of flow $A(\tau)$. Function $f(x)$ denotes the violation probability. It assumes that the stochastic arrival curve $\alpha(\tau)$ may be exceeded by arrival process $A(\tau)$ in sometimes, but the probability of being exceeded is constrained by the boundary function $f(x)$.

Definition 2 (Stochastic Service Curve) A system S is said to provide a stochastic service curve $\beta \in \mathcal{F}$ with bounding function $g \in \bar{\mathcal{F}}$, denoted by $S \sim< g, \beta >$, if for all $t \geq 0$ there holds

$$P\left\{\sup_{0\leq s\leq t} [A \otimes \beta(s) - D(s)] > x\right\} \leq g(x). \tag{4}$$

The symbol $\otimes$ represents the cumulative min-plus convolution operation. Which

$$A \otimes \beta(t) = \inf_{0\leq s\leq t} \{A(s) + \beta(s, t)\} \tag{5}$$

$\beta(t)$ is the stochastic service curve which means the worst service capability provided by the server. Similar to the stochastic arrival curve, the data that already have been processed are probably more than the data departed. The probability of producing exceeding data can be constrained by the boundary function $g(x)$.

Similarly as in (4), the departure process relates to the arrival and service process and it is described as

$$D(t) \geq \inf_{0\leq s\leq t} \{A(s) + S(s, t)\} = A \otimes S(t). \tag{6}$$

Where for all $s, t \geq 0$ and $s \leq t$. That is also the concept of a dynamic server which mentioned in [3]. From the (6), we can better understand the relationship among arrival process, departure process and service process. With these basic processes and curves, we can discuss the definition of the delay boundary.

Definition 3 (Latency Process) Let $A(t)$ and $D(t)$ respectively be the arrival process and departure process. The latency process $L(t)$ at time $t \geq 0$ is defined as

$$L(t) = \inf\{d \geq 0 : A(t) \leq D(t + d)\}. \tag{7}$$

The (7) express that latency $L(t)$ is the least value of d, and the d must meet the condition which the amount of arrival data at time t is less than or equal to the departure data at time $t + d$. It also means that the data do not leave the server immediately. The duration of the data in server is the delay time. In Formula (7), the arrival process $A(t)$ is little than or equal to the departure process $D(t+d)$. It means that the data arrived in server at time t are all leaving from server at time $t + d$. If $A(t)$ is large than or equal to the departure process $D(t+d)$, that represents the data arrived at t moment have not been completed by service during d period of time. So the $A(t)$ little than or equal to $D(t+d)$ situation is utilized to describe the shortest time that server takes for the data to be serviced. That is the latency or delay.

According to the latency process definition, and utilizing stochastic arrival process and stochastic service process, so the stochastic latency bound has been defined at following.

Theorem 1 (Stochastic Latency Bound) *A system with an input process $A(t)$. $A(t)$ is a stochastic arrival process with stochastic arrival curve $\alpha \in \mathcal{F}$ and bounded by function $f \in \bar{\mathcal{F}}$ (i.e., $A \sim< f, \alpha >$). The system provides to the input a stochastic service process $S(t)$. $S(t)$ is with stochastic service curve $\beta \in \mathcal{F}$ with bounding function $g \in \bar{\mathcal{F}}$ (i.e., $S \sim< g, \beta >$). Then, for all $t \geq 0$ and $x \geq 0$, the Latency $L(t)$ is bounded by*

$$P\{L(t) > h(\alpha + x, \beta)\} \leq f \otimes g(x) \tag{8}$$

Where function $h(\alpha + x, \beta)$ denotes the maximum horizontal distance between $\alpha + x$ and β, the express $f \otimes g(x)$ represents the cumulative min-plus convolution operation of function f and g.

Theorem 2 (Concatenation Property) *Considering a flow passes through a network of N server nodes in tandem. If each server nodes $n(= 1, 2, \ldots, N)$ provides a stochastic service curve $S^n \sim< g^n, \beta^n >$ to its input, then the network guarantees to the flow a stochastic service curve $S \sim< g, \beta >$ with*

$$\begin{aligned} \beta(t) &= \beta^1 \otimes \beta^2 \otimes \cdots \otimes \beta^N(t) \\ g(x) &= g^1 \otimes g^2 \otimes \cdots \otimes g^N(t) \end{aligned} \tag{9}$$

3.5 *Model Building*

The network character of URLLC can be described as a dynamic server by stochastic processes as introducing in above. The data sent by UE can be represented by the arrival process $A(t)$. The service capacity provided by the network server node can be depicted as the stochastic service process $S(t)$.

As assumption the latency of URLLC in Formula (1), we consider the URLLC network is a tandem system. So the delay of URLLC should fall in the concatenation characterization in SNC.

We consider that a UE's network flow passing through the gNB, CN and Cloud in tandem mode. Each network node $k(= gNB, AAU, DU, CU, CN, Cloud)$ provides a stochastic service curve $S_k \sim< g_k, \beta_k >$ to its flow. We first discuss about the gNB subsystem. The gNB includes AAU, DU and CU, so $S_{AAU}(s,t)$, $S_{DU}(s,t)$ and $S_{CU}(s,t)$ are in series. We use the same indices to denote the arrival and departure process of the respective systems. Especially the source of data is from UE. So $A_{AAU}(t)$ as arrival process denotes the input data of AAU from

UE in gNB subsystem. The arrival process of DU $A_{DU}(t)$ actually equals to the departure process of AAU $D_{AAU}(t)$, where $A_{DU}(t) = D_{AAU}(t)$. By the same token, $A_{CU}(t) = D_{DU}(t)$ similarly for CU server. The departure process of CU is $D_{CU}(t)$. It is also the departure process of the gNB subsystem.

Considering that the gNB subsystem is tandem deployed, we assume that the AAU server provides a $S_{AAU}(t)$ capacity to deal with the arrival data. Applying (6), the departure process can be represent as

$$D_{AAU}(t) \geq A_{AAU} \otimes S_{AAU}(t). \tag{10}$$

Similar to the departure process of AAU, we can get the process of DU

$$D_{DU}(t) \geq A_{DU} \otimes S_{DU}(t). \tag{11}$$

Because of $A_{DU}(t) = D_{AAU}(t) = A_{AAU} \otimes S_{AAU}(t)$, we put (10) into (11) to replace $A_{DU}(t)$ by $A_{AAU} \otimes S_{AAU}(t)$ and get

$$D_{DU}(t) \geq (A_{AAU} \otimes S_{AAU}(t)) \otimes S_{DU}(t). \tag{12}$$

By recursive insertion, we can obtain

$$D_{CU}(t) \geq ((A_{AAU} \otimes S_{AAU}) \otimes S_{DU}) \otimes S_{CU}(t). \tag{13}$$

Applying the associativity of min-plus convolution, it holds that

$$\begin{aligned} D_{CU}(t) &\geq ((A_{AAU} \otimes S_{AAU}) \otimes S_{DU}) \otimes S_{CU}(t) \\ &\geq A_{AAU} \otimes (S_{AAU} \otimes S_{DU} \otimes S_{CU})(t). \end{aligned} \tag{14}$$

From the gNB subsystem perspective to see, $A_{AAU}(t)$ is the first input and $D_{CU}(t)$ is the last output of gNB. So $A_{AAU}(t)$ equal to $A_{gNB}(t)$, and $D_{CU}(t)$ is the departure process of the gNB $D_{gNB}(t)$. Then we can get gNB subsystem

$$D_{gNB} \geq A_{AAU} \otimes (S_{AAU} \otimes S_{DU} \otimes S_{CU})(t). \tag{15}$$

Assuming first-come first-served order, we use Definition 3 and Eq. (15), and let L_{gNB} denotes the latency process of gNB, there holds

$$\begin{aligned} L_{gNB}(t) &= \inf\{d \geqslant 0 : A_{gNB}(t) - D_{gNB}(t+d) \leqslant 0\} \\ &= \inf\{d \geqslant 0 : A_{AAU}(t) - D_{CU}(t+d) \leqslant 0\} \\ &= \inf\{d \geqslant 0 : A_{AAU}(t) \\ &- A_{AAU} \otimes (S_{AAU} \otimes S_{DU} \otimes S_{CU})(t+d) \leqslant 0\} \end{aligned} \tag{16}$$

With Theorems 1 and 2, the delay bound can be analysis by following corollary.

Corollary 1 (Latency Bound of gNB) *In gNB subsystem, $A_{AAU}(t)$ is a stochastic arrival process with stochastic arrival curve α_{AAU}, i.e. $A \sim< f_{AAU}, \alpha_{AAU} >$. $\alpha_{AAU} \in \mathcal{F}$, $f_{AAU} \in \bar{\mathcal{F}}$. The server nodes in subsystem provide stochastic service process $S_{AAU}(t), S_{DU}(t), S_{CU}(t)$ respectively, i.e. $S_{AAU} \sim< g_{AAU}, \beta_{AAU} >$, $S_{DU} \sim< g_{DU}, \beta_{DU} >, S_{CU} \sim< g_{CU}, \beta_{CU} >$. And $\beta_{AAU}, \beta_{DU}, \beta_{CU} \in \mathcal{F}$, $g_{AAU}, g_{DU}, g_{CU} \in \bar{\mathcal{F}}$. Then, for all $t \geq 0$ and $x \geq 0$, the Latency of gNB subsystem $L_{gNB}(t)$ is bounded by*

$$\begin{aligned} P\{L_{gNB}(t) \geq d\} &= P\{L_{gNB}(t) \geq h(\alpha_{AUU} + x, \beta_{gNB})\} \\ &\leqslant f_{AAU} \otimes g_{gNB}(x) \end{aligned} \tag{17}$$

where service rate $\beta_{gNB}(t) = \beta_{AAU} \otimes \beta_{DU} \otimes \beta_{CU}(t)$, and bound function $g_{gNB} = g_{AAU} \otimes g_{DU} \otimes g_{CU}(x)$.

Proof Please see Appendix 1.

For mobile edge computing (MEC) deployment, CU can be the end of the transmission. That because the computing resource and storage resource are located at CU. The Corollary 1 is suitable for analysis the delay of communication from UE to CU. However, in order to comprehensively discuss the delay of URLLC system, we need to convert the destination from CU to Cloud. By extending Corollary 1, we can draw the whole URLLC system latency bound.

Corollary 2 (Latency Bound of URLLC) *In 5G URLLC system, $A_{AAU}(t)$ is a stochastic arrival process with stochastic arrival curve α_{AAU}, i.e. $A \sim< f_{AAU}, \alpha_{gNB} >$. $\alpha_{AAU} \in \mathcal{F}$, $f_{AAU} \in \bar{\mathcal{F}}$. The server nodes in URLLC system provide stochastic service process $S_{gNB}(t)$, $S_{NGC}(t)$ and $S_{Cloud}(t)$ respectively, i.e. $S_{gNB} \sim< g_{gNB}, \beta_{gNB} >$, $S_{NGC} \sim< g_{NGC}, \beta_{NGC} >$, and $S_{Cloud} \sim< g_{Cloud}, \beta_{Cloud} >$. Service rate $\beta_{gNB}, \beta_{NGC}, \beta_{Cloud} \in \mathcal{F}$, $g_{gNB}, g_{NGC}, g_{Cloud} \in \bar{\mathcal{F}}$. Then, for all $t \geq 0$ and $x \geq 0$, the Latency of URLLC system $L_{All}(t)$ is bounded by*

$$\begin{aligned} P\{L_{All}(t) \geq d\} &= P\{L_{All}(t) \geq h(\alpha_{AUU} + x, \beta_{All})\} \\ &\leqslant f_{AAU} \otimes g_{All}(x) \end{aligned} \tag{18}$$

where $\beta_{All}(t) = \beta_{gNB} \otimes \beta_{NGC} \otimes \beta_{Cloud}(t)$, and $g_{All} = g_{gNB} \otimes g_{NGC} \otimes g_{Cloud}(x)$.

Proof Please see Appendix 2. We have built a model to represent the latency of 5G MEC (UE to gNB) and URLLC (the full path from UE to Cloud). Next we intend to calculate the delay boundary of the model. With Corollary 2, we know that four key variance need to be determined. These are stochastic arrival process α_{AAU}, f_{AAU}, and stochastic service process β_{All}, g_{All}. Especially, we can also decompose β_{All} and g_{All} to obtain more detailed result.

In URLLC scenario, the data is usually fixed unit packet size, the data size sometimes very tiny while applying millimeter-wave technology. We assume that UE data arrive will approximate to a Poisson distribution with mean rate λ. So arrive curve $\alpha_{AAU} = \lambda t$ and bound function will be

$$f_{AAU}(x) = \sum_{k=x+\lambda t}^{\infty} \frac{e^{\lambda t} \cdot (\lambda t)^k}{k!} \tag{19}$$

With MGF of right hand in (19) and Chernoff bound, f_{AAU} can be tighten by

$$f_{AAU}(x) = e^{x-(\lambda t+x)ln\frac{\lambda t+x}{\lambda t}} \tag{20}$$

The proof of this part can be found in [23]. Two variances in stochastic arrival process have been solved. We begin to determine β_{All}, g_{All} for the stochastic service process.

In order to simplify the problem, we generalize service rate of server nodes and assume that each node provides data processing capacity as $\beta(t) = Ct$ with bounding function $g(x) = ae^{-bx}$. According to Corollary 2, we can get

$$g_{All}(x) = g_{AAU} \otimes g_{DU} \otimes g_{CU} \otimes g_{NGC} \otimes g_{Cloud}(x) \tag{21}$$

Therefore,

$$\begin{aligned} g_{All}(x) &= g_{AAU} \otimes g_{DU} \otimes g_{CU} \otimes g_{NGC} \otimes g_{Cloud}(x) \\ &= \inf_{x_1+x_2+x_3+x_4+x_5=x} \sum_{k=1}^{5} a_k e^{-b_k x_k} \end{aligned} \tag{22}$$

Applying with the conclusion in [24], we can hold

$$\inf_{x_1+x_2+x_3+x_4+x_5=x} \sum_{k=1}^{5} a_k e^{-b_k x_k} = e^{\frac{-x}{w}} \prod_{k=1}^{5} (a_k b_k w)^{\frac{1}{b_k w}} \tag{23}$$

where $w = \sum_{k=1}^{5} \frac{1}{b_k}$, and service bound functions respectively are $g_{AAU}(x) = a_1 e^{-b_1 x_1}$, $g_{DU}(x) = a_2 e^{-b_2 x_2}$, $g_{CU}(x) = a_3 e^{-b_3 x_3}$, $g_{NGC}(x) = a_4 e^{-b_4 x_4}$, $g_{Cloud}(x) = a_5 e^{-b_5 x_5}$. with all the information we discuss above, and applying the lemma which proved in [24], we can get

$$g_{All}(x) = e^{\frac{-xb}{n+1}}(a(n+1)) \tag{24}$$

We put (20), (24) into (18) and apply the Theorem 3 proved in [25], it can be derived that

$$\begin{aligned} &P\{L(t) > h(\alpha_{AAU} + x, \beta)\} \\ &= P\{L(t) > \frac{x}{C-\lambda}\} \\ &\leq e^{x-(\lambda t+x)ln\frac{\lambda t+x}{\lambda t}} \cdot e^{\frac{-xb}{n+1}}(a(n+1)) \end{aligned} \tag{25}$$

Then we hold the latency bound as (2). Let $d = \frac{x}{C-\lambda}$ and set the right side of (25) equal to ϵ. The ϵ is a small latency bound violation probability. We can obtain a relationship between d and ϵ.

$$d = \frac{1}{C-\lambda} \cdot \frac{n+1}{b} \cdot ln\frac{a(n+1)}{\epsilon} \tag{26}$$

The calculation process can be found in Appendix 3.

4 Numerical Results and Performance Evaluation

In this section, we will discuss what factors are the main cause of latency in URLLC.

Although the deployment details of the URLLC standard are not yet released, we can still apply SNC theory for quantitative analysis. We assume that the 5G URLLC networks are standalone deployment. The packet arrival rate is constant and arrival process satisfies Poisson distribution. A general URLLC reliability requirement for one transmission of a packet is $1 * 10^{-5}$ for 32 bytes with a user plane latency of 1 ms. So we set the violation probability value around $1 * 10^{-5}$. More simulation parameters can be found in Table 1.

Taking this boundary probability as the precondition, we simulate the relationship between system latency and service rate by applying the conclusions we have

Table 1 Evaluation parameters

Parameter	Value
Network deployment	Standalone
Traffic mode	Constant transmission, Poisson arrival
Arrival rate λ	20 (Gbit/s)
Service rate C	40, 45, 50, 55 (Gbit/s)
Service bound a	1
Service bound b	3
Number of tandem servers n	5
Violation probability	$1 * 10^{-5}$
Latency bound	1 (ms)

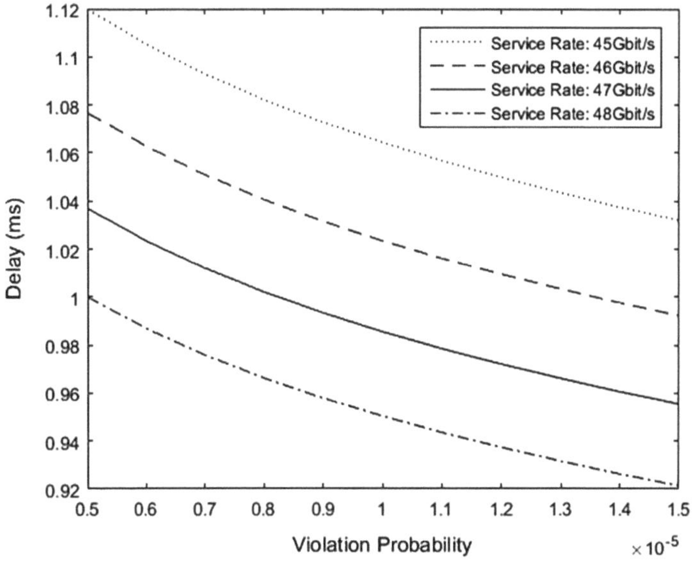

Fig. 2 Service rate influence

drawn in the previous section. Figure 2 provide the evaluated URLLC delay with different service rate under violation probability $1 * 10^{-5}$. The value of violation probability is from $5 * 10^{-6}$ to $15 * 10^{-6}$. We arrange the value scope to include the demand value $1 * 10^{-5}$ to observe the effect of this value on delay. We adopt four service rates in model and all the curves are slow down by violation probability value. From this, we can conduct that the violation probability is not the main factor to influence the latency. In order to make the delay less than 1 ms, we set service rate from 45 Gbit/s to 48 Gbit/s based on arrival rate 20 Gbit/s. We can procure that delay approximates 1 ms when service rate is 47 Gbit/s at violation probability $1 * 10^{-5}$. As the service rate increases, the delay of the system will decrease. When service rate is 48 Gbit/s, system latency can approach 1 ms with lower violation probability. That means system can guarantee the low latency communication in a stable state.

Figure 3 presents the relationship among latency, number of server levels and service rate. The arrival rate is constant and the speed is 20 Gbit/s. The violation probability is maintained at $1 * 10^{-5}$. Based on the above setting, we can derived that the delay is sensitive on number of tandem servers. From Fig. 3, we can see the slope of latency caused by number of tandem servers is larger than the service rate. We draw a delay equals 1 ms flat plane to cut the curved surface. The part below the plane is the scope of deployment parameters which satisfying the delay condition. In order to ensure low latency of communication, it is necessary to reduce the number of tandem servers deployment as much as possible and increase the service rate of each server layer.

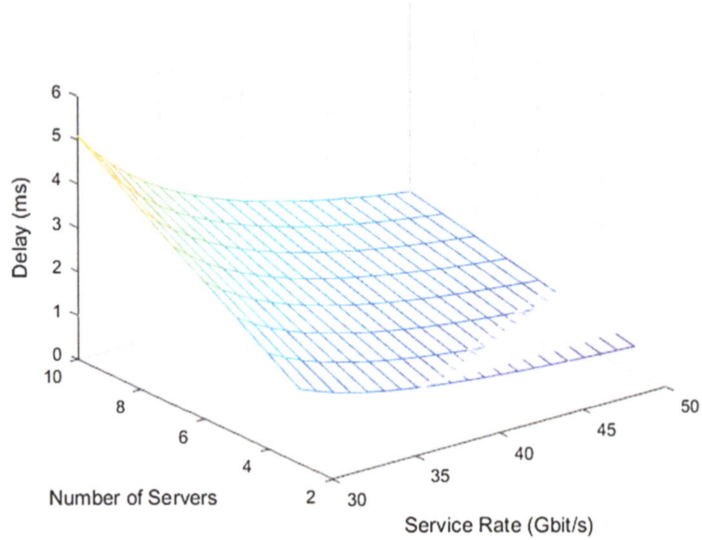

Fig. 3 Number of tandem servers influence

5 Conclusions

In this paper, the architecture of 5G URLLC network is researched. According to the architecture characteristics, the URLLC network is modeled as a tandem system which describes the communication from UE to Cloud. Applying stochastic network theory and combining the features of URLLC network, performance analysis has been conducted. We have investigated the relationship between delay constraints, service rates, violation probabilities and the number of deployed servers in URLLC networks. The 3GPP standard is taken into account when we set the simulation parameters. Numerical results verify that the main factor which can impact on latency is the number of servers deployed in tandem. That also means Edge Computing will be the trend in URLLC application deployment. The service rate of the server is also a factor affecting the delay. With the increase of service rate, delay can be reduced. The results derived from evaluation provide valuable guidelines for the early design of URLLC deployment. For our future work, we would consider to include handover access in URLLC communication.

Acknowledgements Supported by program National Natural Science Foundation of China. (Nos.61872044,61502040). Beijing Municipal Program for Excellent Teacher Promotion. (No.PXM2017_014224.000028).

Appendix 1: Proof of Corollary 1

Proof Since the latency process Definition 3 are defined as $L(t) = \inf\{d \geq 0 : A(t) \leq D(t+d)\}$, event $L(t) > d$ implies event $A(t) \leq D(t+d)$. We move $D(t+d)$ from right hand to left hand, and according to (16), the latency bound of gNB can be hold as

$$P\{L_{gNB}(t) > d\} \leqslant P\{A_{AAU}(t) - D_{CU}(t+d) \leq 0\} \tag{27}$$

Then we focus on the $\{A_{AAU}(t) - D_{CU}(t+d)\}$ part. We put right hand of (15) into this part, we can get

$$\begin{aligned} &A_{AAU}(t) - D_{CU}(t+d) \\ =&A_{AAU}(t) - A_{AAU} \otimes (S_{AAU} \otimes S_{DU} \otimes S_{CU})(t+d) \\ +&A_{AAU} \otimes (S_{AAU} \otimes S_{DU} \otimes S_{CU})(t+d) - D_{CU}(t+d) \end{aligned} \tag{28}$$

With the Theorem 2, utilizing the concatenation property we can obtain that $S_{AAU} \sim< g_{AAU}, \beta_{AAU} >$, $S_{DU} \sim< g_{DU}, \beta_{DU} >$, $S_{CU} \sim< g_{CU}, \beta_{CU} >$. Stochastic service process convolution operation $(S_{AAU} \otimes S_{DU} \otimes S_{CU})$ means gNB subsystem provides maybe lower than $(\beta_{AAU} \otimes \beta_{DU} \otimes \beta_{CU})$ processing capacity, but the violation probability in this case is limited by $g_{AAU} \otimes g_{DU} \otimes g_{DU}$. Through applying (4), we denote β_{gNB} equals to $(\beta_{AAU} \otimes \beta_{DU} \otimes \beta_{CU})$, g_{gNB} equals to $g_{AAU} \otimes g_{DU} \otimes g_{DU}$. hence (28) can hold be

$$\begin{aligned} &A_{AAU}(t) - D_{CU}(t+d) \\ =&A_{AAU}(t) - A_{AAU} \otimes (\beta_{AAU} \otimes \beta_{DU} \otimes \beta_{CU})(t+d) \\ &+ A_{AAU} \otimes (\beta_{AAU} \otimes \beta_{DU} \otimes \beta_{CU})(t+d) - D_{CU}(t+d) \\ =&A_{AAU}(t) - A_{AAU} \otimes \beta_{gNB}(t+d) \\ &+ A_{AAU} \otimes \beta_{gNB}(t+d) - D_{CU}(t+d) \end{aligned} \tag{29}$$

According to (5), we replace $A_{AAU} \otimes \beta_{gNB}(t+d)$ by $\inf\{A_{AAU}(s)+\beta_{gNB}(t+d-s)\}$ in (29). Consequently,

$$\begin{aligned} &A_{AAU}(t) - D_{CU}(t+d) \\ =&A_{AAU}(t) - A_{AAU} \otimes \beta_{gNB}(t+d) \\ &+ A_{AAU} \otimes \beta_{gNB}(t+d) - D_{CU}(t+d) \\ =&A_{AAU}(t) - \inf_{0 \leqslant s \leqslant t+d} \{A_{AAU}(s) + \beta_{gNB}(t+d-s)\} \\ &+ A_{AAU} \otimes \beta_{gNB}(t+d) - D_{CU}(t+d) \end{aligned} \tag{30}$$

$$
\begin{aligned}
\leq & A_{AAU}(t) - A_{AAU}(s) - \beta_{gNB}(t+d-s)\} \\
& + A_{AAU} \otimes \beta_{gNB}(t+d) - D_{CU}(t+d) \\
\leq & A_{AAU}(s,t) - \alpha_{AAU}(t-s) \\
& + \alpha_{AAU}(s,t) - \beta_{gNB}(t+d-s) \\
& + A_{AAU} \otimes \beta_{gNB}(t+d) - D_{CU}(t+d)
\end{aligned}
$$

We add α_{AAU} at step 4 in (30) to build stochastic arrival curve. Based on the stochastic arrival curve (3),$A_{AAU}(s,t) - \alpha_{AAU}(s,t)$ is less than or equal to f_{AAU}. Applying stochastic service curve (4), $A_{AAU} \otimes \beta_{gNB}(t+d) - D_{CU}(t+d)$ is less than or equal to g_{gNB}. With Theorem 1, we use $h(\alpha + x, \beta)$ replace the d. where $h(\alpha + x, \beta)$ is the maximum horizontal distance between $\alpha + x$ and β for $x \geq 0$. The $h(\alpha, \beta)$ function implies the condition

$$
\lim_{t\to\infty} [\alpha(t) - \beta(t)] \leq 0. \tag{31}
$$

we can obtain

$$
\begin{aligned}
& P\{L(t) > h(\alpha_{AAU} + x, \beta_{gNB})\} \\
& = P\{\{A_{AAU}(t) - D_{CU}(t + h(\alpha + x, \beta))\} > 0\} \\
& \leqslant \sup_{0\leqslant s\leqslant t} \{A_{AAU}(s,t) - \alpha_{AAU}(t-s)\} \\
& + \sup_{0\leqslant s\leqslant t+h(\alpha_{AAU}+x,\beta_{gNB})} \{A_{AAU} \otimes \beta_{gNB}(s) - D_{CU}(s)\} \\
& \leqslant f_{AAU}(t) + g_{gNB}(x) \\
& \leqslant \inf\{f_{AAU}(t) + g_{gNB}(x-t)\} \\
& \leqslant f_{AAU} \otimes g_{gNB}(x)
\end{aligned}
$$

Therefore, Corollary 1 is proved.

Appendix 2: Proof of Corollary 2

Proof In the Corollary 1, the gNB subsystem are constituted by AAU, DU and CU. In addition to gNB, the whole 5G URLLC system also include NGC and Cloud. According to the concatenation property which mentioned in Theorem 2, then the network guarantees to the flow a stochastic service curve $S_{All} \sim< g_{All}, \beta_{All} >$ with

$$
\beta_{All}(t) = \beta_{gNB} \otimes \beta_{CN} \otimes \beta_{Cloud}(t) \tag{32}
$$

where

$$\beta_{gNB}(t) = \beta_{AAU} \otimes \beta_{DU} \otimes \beta_{CU}(t) \tag{33}$$

actually

$$\beta_{All}(t) = \beta_{AAU} \otimes \beta_{DU} \otimes \beta_{CU} \otimes \beta_{CN} \otimes \beta_{Cloud}(t) \tag{34}$$

and

$$g_{All}(x) = g_{gNB} \otimes g_{CN} \otimes g_{Cloud}(x) \tag{35}$$

where

$$g_{gNB}(x) = g_{AAU} \otimes g_{DU} \otimes g_{CU}(x) \tag{36}$$

actually

$$g_{All}(x) = g_{AAU} \otimes g_{DU} \otimes g_{CU} \otimes g_{CN} \otimes g_{Cloud}(x) \tag{37}$$

Based on latency process Definition 3, the 5G URLLC system latency process can be defined as

$$L(t) = \inf\{d \geqslant 0 : A_{AAU}(t) \leq D_{Cloud}(t + d)\} \tag{38}$$

Latency bound of 5G URLLC is defined as

$$P\{L(t) \geq d\} = P\{A_{AAU}(t) - D_{Cloud}(t + d) \leq 0\} \tag{39}$$

We also focus on $A_{AAU}(t) - D_{Cloud}(t+d)$ part. where $D_{Cloud} \geqslant A_{AAU} \otimes S_{AAU} \otimes S_{DU} \otimes S_{CU} \otimes S_{NGC} \otimes S_{Cloud}$. Then we have

$$\begin{aligned}
&A_{AAU}(t) - D_{Cloud}(t + d) \\
=&A_{AAU}(t) - A_{AAU} \otimes S_{AAU} \otimes S_{DU} \otimes S_{CU} \\
&\otimes S_{NGC} \otimes S_{Cloud}(t + d) \\
&+ A_{AAU} \otimes S_{AAU} \otimes S_{DU} \otimes S_{CU} \\
&\otimes S_{NGC} \otimes S_{Cloud} - D_{Cloud}(t + d) \\
=&A_{AAU}(t) - A_{AAU} \otimes S_{gNB} \otimes S_{NGC} \otimes S_{Cloud}(t + d) \\
&+ A_{AAU} \otimes S_{gNB} \otimes S_{NGC} \otimes S_{Cloud} - D_{Cloud}(t + d) \\
\leq&A_{AAU}(t) - A_{AAU}(s) \\
&- \beta_{gNB} \otimes \beta_{NGC} \otimes \beta_{Cloud}(t + d - s))
\end{aligned}$$

$$
\begin{aligned}
&+ A_{AAU} \otimes S_{gNB} \otimes S_{NGC} \otimes S_{Cloud} - D_{Cloud}(t+d) \\
\leq & A_{AAU}(s,t) - \alpha_{AAU}(t-s) \\
&+ \alpha_{AAU}(t-s) - \beta_{all}(t+d-s) \\
&+ A_{AAU} \otimes \beta_{all}(t+d) - D_{Cloud}(t+d)
\end{aligned}
$$

With stochastic arrival curve (3), $A_{AAU}(s,t) - \alpha(t-s)$ is bounded by $f_{AAU}(x)$. According to stochastic service curve (4), $A_{AAU} \otimes \beta_{All}(t+d) - D_{Cloud}(t+d)$ is limited by g_{All}. For long-term running, if $t \to \infty$, $\alpha_{AAU}(t-s) - \beta_{All}(t+d-s)$ approximate to zero because of $\alpha_{AAU}, \beta_{All} \in \mathcal{F}$. Finally, with Theorem 1, the delay of the URLLC system can be bounded by this

$$
P\{L(x) > h(\alpha_{AAU} + x, \beta_{All})\} < f_{AUU} \otimes g_{All}(x) \tag{40}
$$

Therefore, Corollary 2 is proved.

Appendix 3: Calculation of Delay

We first set right side of (25) equals to ϵ, and we logarithm on both sides then hold

$$
\begin{aligned}
e^{x-(\lambda t+x)ln\frac{\lambda t+x}{\lambda t}} \cdot e^{\frac{-xb}{n+1}}(a(n+1)) &= \epsilon \\
e^{x-(\lambda t+x)ln\frac{\lambda t+x}{\lambda t}} \cdot e^{\frac{-xb}{n+1}} &= \frac{\epsilon}{a(n+1)}
\end{aligned} \tag{41}
$$

for a long-term running situation, $t \to \infty$, then

$$
\lim_{t\to\infty} (\lambda t + x)ln\frac{\lambda t + x}{\lambda t} = x \tag{42}
$$

then the (41) equal to

$$
\begin{aligned}
e^{\frac{-xb}{n+1}} &= \frac{\epsilon}{(a(n+1))} \\
\frac{-xb}{n+1} &= ln\frac{\epsilon}{(a(n+1))} \\
x &= \frac{n+1}{b} \cdot ln\frac{a(n+1)}{\epsilon}
\end{aligned} \tag{43}
$$

we put $x = d(C - \lambda)$ into (43) and get

$$
d = \frac{1}{C-\lambda} \cdot \frac{n+1}{b} \cdot ln\frac{a(n+1)}{\epsilon} \tag{44}
$$

Therefore d is solved.

References

1. ITU-R M.2083-0.: IMT vision - framework and overall objectives of the future development of IMT for 2020 and beyond (2015)
2. 3GPP TR 38.913.: Study on scenarios and requirements for next generation access technologies (2017)
3. Soldani, D., Guo, Y.J., Barani, B., et al.: 5G for ultra-reliable low-latency communications. IEEE Netw. **32**(2), 6–7 (2018)
4. Jiang, Y., Liu, Y.: Stochastic network calculus. Springer, London (2009)
5. Fidler, M., Rizk, A.: A guide to the stochastic network calculus. IEEE Commun. Surv. Tutorials **17**(1), 92–105 (2015)
6. Nielsen, J.J., Liu, R., Popovski, P.: Ultra-reliable low latency communication using interface diversity. IEEE Trans. Commun. **66**(3), 1322–1334 (2018)
7. Delgado, R.A., Lau, K., Middleton, R.H., et al.: Networked delay control for 5G wireless machine-type communications using multiconnectivity. IEEE Trans. Control Syst. Technol. **99**, 1–16 (2018)
8. Rao, J., Vrzic, S.: Packet duplication for URLLC in 5G: architectural enhancements and performance analysis. IEEE Netw. **32**(2), 32–40 (2018)
9. Anand, A., De Veciana, G.: Resource allocation and HARQ optimization for URLLC traffic in 5G wireless networks (2018)
10. Mukherjee, A.: Energy efficiency and delay in 5G ultra-reliable low-latency communications system architectures. IEEE Netw. **32**(2), 55–61 (2018)
11. Sachs, J., Wikstrom, G., Dudda, T., et al.: 5G radio network design for ultra-reliable low-latency communication. IEEE Netw. **32**(2), 24–31 (2018)
12. Wang, C., Chen, Y., Wu, Y., et al.: Performance evaluation of grant-free transmission for uplink URLLC services. In: IEEE Vehicular Technology Conference 2017 Vtc2017-Spring, pp. 1–6 (2017)
13. Pocovi, G., Shariatmadari, H., Berardinelli, G., et al.: Achieving ultra-reliable low-latency communications: challenges and envisioned system enhancements. IEEE Netw. **32**(2), 8–15 (2018)
14. Popovski, P., Nielsen, J.J., Stefanovic, C., et al.: Wireless access for ultra-reliable low-latency communication: principles and building blocks. IEEE Netw. **32**(2), 16–23 (2018)
15. Ji, H., Park, S., Yeo, J., et al.: Introduction to ultra reliable and low latency communications in 5G (2017)
16. Ji H, Park S, Yeo J, et al.: Ultra Reliable and Low Latency Communications in 5G Downlink: Physical Layer Aspects. (2018)
17. Li, Z., Jiang, Y., Gao, Y., Li, P., Sang, L., Yang, D.: Delay and delay-constrained throughput performance of a wireless-powered communication system. IEEE Access **5**, 21620–21631 (2017)
18. Lbben, R., Fidler, M.: Estimation method for the delay performance of closed-loop flow control with application to TCP. In: IEEE INFOCOM 2016-The 35th Annual IEEE International Conference on Computer Communications, San Francisco, CA , pp. 1–9 (2016)
19. Zheng, K., Liu, F., Lei, L., Lin, C., Jiang, Y.: Stochastic performance analysis of a wireless finite-state Markov channel. IEEE Trans. Wireless Commun. **12**(2), 782–793 (2013)
20. Chen, X., Si, Y., Xiang, X.: Delay-bounded resource allocation for femtocells exploiting the statistical multiplexing gain, vol. 71, pp. 3217–3236. Kluwer Academic Publishers, Dordrecht (2015). https://doi.org/10.1007/s11227-015-1494-9
21. Fidler, M., Walker, B., Jiang, Y.: Non-asymptotic delay bounds for multi-server systems with synchronization constraints. IEEE Trans. Parallel Distrib. Syst. **29**(7), 1545–1559 (2018)
22. Lbben, R., Fidler, M., Liebeherr, J.: Stochastic bandwidth estimation in networks with random service. IEEE/ACM Trans. Netw. **22**(2), 484–497 (2014)
23. Wu, K., Jiang, Y., Li, J.: On the model transform in stochastic network calculus. In: International Workshop on Quality of Service IEEE, pp. 1–9 (2010)

24. Sun, F., Li, L., Jiang, Y.: Impact of duty cycle on end-to-end performance in a wireless sensor network. In: Wireless Communications and Networking Conference IEEE, pp. 1906–1911 (2015)
25. Beck, M.: Towards the analysis of transient phases with stochastic network calculus. In: Telecommunications Network Strategy and Planning Symposium IEEE, pp. 164–169 (2016)

Index

B. Ye et al. (eds.), *2nd International Conference on 5G for Ubiquitous Connectivity*, EAI/Springer Innovations in Communication and Computing,
https://doi.org/10.1007/978-3-030-22316-8

MIX
Papier aus verantwortungsvollen Quellen
Paper from responsible sources
FSC® C105338
FSC
www.fsc.org

If you have any concerns about our products,
you can contact us on
ProductSafety@springernature.com

In case Publisher is established outside the EU,
the EU authorized representative is:
Springer Nature Customer Service Center GmbH
Europaplatz 3, 69115 Heidelberg, Germany

Printed by Libri Plureos GmbH
in Hamburg, Germany